Darius V. Köster

Role of Caveolae in Membrane Mechanics

Darius V. Köster

Role of Caveolae in Membrane Mechanics

Südwestdeutscher Verlag für Hochschulschriften

Impressum/Imprint (nur für Deutschland/only for Germany)
Bibliografische Information der Deutschen Nationalbibliothek: Die Deutsche Nationalbibliothek verzeichnet diese Publikation in der Deutschen Nationalbibliografie; detaillierte bibliografische Daten sind im Internet über http://dnb.d-nb.de abrufbar.
Alle in diesem Buch genannten Marken und Produktnamen unterliegen warenzeichen-, marken- oder patentrechtlichem Schutz bzw. sind Warenzeichen oder eingetragene Warenzeichen der jeweiligen Inhaber. Die Wiedergabe von Marken, Produktnamen, Gebrauchsnamen, Handelsnamen, Warenbezeichnungen u.s.w. in diesem Werk berechtigt auch ohne besondere Kennzeichnung nicht zu der Annahme, dass solche Namen im Sinne der Warenzeichen- und Markenschutzgesetzgebung als frei zu betrachten wären und daher von jedermann benutzt werden dürften.

Verlag: Südwestdeutscher Verlag für Hochschulschriften GmbH & Co. KG
Dudweiler Landstr. 99, 66123 Saarbrücken, Deutschland
Telefon +49 681 37 20 271-1, Telefax +49 681 37 20 271-0
Email: info@svh-verlag.de

Approved by: Paris, UPMC, Diss., 2010

Herstellung in Deutschland:
Schaltungsdienst Lange o.H.G., Berlin
Books on Demand GmbH, Norderstedt
Reha GmbH, Saarbrücken
Amazon Distribution GmbH, Leipzig
ISBN: 978-3-8381-2712-5

Imprint (only for USA, GB)
Bibliographic information published by the Deutsche Nationalbibliothek: The Deutsche Nationalbibliothek lists this publication in the Deutsche Nationalbibliografie; detailed bibliographic data are available in the Internet at http://dnb.d-nb.de.
Any brand names and product names mentioned in this book are subject to trademark, brand or patent protection and are trademarks or registered trademarks of their respective holders. The use of brand names, product names, common names, trade names, product descriptions etc. even without a particular marking in this works is in no way to be construed to mean that such names may be regarded as unrestricted in respect of trademark and brand protection legislation and could thus be used by anyone.

Publisher: Südwestdeutscher Verlag für Hochschulschriften GmbH & Co. KG
Dudweiler Landstr. 99, 66123 Saarbrücken, Germany
Phone +49 681 37 20 271-1, Fax +49 681 37 20 271-0
Email: info@svh-verlag.de

Printed in the U.S.A.
Printed in the U.K. by (see last page)
ISBN: 978-3-8381-2712-5

Copyright © 2011 by the author and Südwestdeutscher Verlag für Hochschulschriften GmbH & Co. KG and licensors
All rights reserved. Saarbrücken 2011

Before starting with the science part, I would like to express my gratitude to those, who made this work possible. First of all, there are my supervisors Prof. Dr. J. Käs, Dr. P. Nassoy and Dr. C. Lamaze, who introduced me to the beautiful and interesting field of Bio-Physics, and who accompanied me with good advices through all phases of my PhD project. A special thank goes to Prof. F. Brochard-Wyart, who rendered it possible for me to come to the Institut Curie in forwarding my request for an internship to Dr. P. Nassoy. In total, I have spent five years at the Institut Curie in Paris, and I found there not only good colleagues, but also many lovely friends, who I want to thank in the following. There are my colleagues of the Bassereau group, the "Physicists", Alessia, Bidisha, Carina, Mahassine, Marie, Patricia, Pia, Sandrine, Sophie, Aurélien, Benoît, Bibhu, Clément, Ernesto, Faris, Gerbrand, Gilman, John, Kevin, Ludwig, Pierre, Saleem, Thomas and Yegor, as well as my colleagues from the Johannes/Lamaze group, the Biologists, Christine, Emelie, Estelle, Julie, Marianne, M arine, Na, Ramya, Sabrina, Siau, Valérie, Yoko, Alexander, Bahne, Cédric, Christophe, Getao, Gustaf, Ludger, Keith, Michael, Richard, Romain C., Romain P., Vincent and Winfried.

The open minded culture at the Institut Curie allowed me to meet and to discuss with many people from other groups and departments, who I want to thank, Ellen, Jenny, Joanne, Karine, Maria-Isabel, Damien, Elyes, Felix, Johannes, Julien, Padram, Paolo, Reza and Timo, and special thanks go to my colleagues of ADIC (the young researchers organization of the Institut Curie), Ana-Joaquina, Kristine, Lea-Laetitia, Maria, Sandrine, Sophie C., Aurèle, Benoît M., Florent, Maxime, Simon.

I appreciated and want to thank especially my colleague Bidisha Sinha for all the discussions and the nice work that we did together.

I also want to thank my teacher at high school, Mr. Wonneberg, without whom I never would have tought to study Physics. Finally, I would like to thank my parents who enabled me to follow my studies in Leipzig and to continue with the Ph.D. in Paris. Without their love, encouragement and support, I would never have been able to do all that.

Abstract

Caveolae, the characteristic plasma membrane invaginations present in many cells, have been associated with numerous functions that still remain debated. Taking into account the particular abundance of caveolae in cells experiencing mechanical stress, it was proposed that caveolae constitute a membrane reservoir and buffer the membrane tension upon mechanical stress. The present work aimed to check this proposition experimentally. First, the influence of caveolae on the membrane tension was studied on mouse lung endothelial cells in resting conditions using tether extraction with optically trapped beads. Second, experiments on cells upon acute mechanical stress showed that caveolae serve as a membrane reservoir buffering surges in membrane tension in their immediate, ATP- and cytoskeleton-independent flattening and disassembly. Third, caveolae incorporated in membrane vesicles also showed the tension buffering. Finally, in a physiologically more relevant case, human muscle cells were studied, and it was shown that mutations with impaired caveolae which are described in muscular dystrophies render muscle cells less resistant to mechanical stress. In Summary the present work provides experimental evidence for the hypothesis that caveolae buffer the membrane tension upon mechanical stress. The fact that this was observed in cells and membrane vesicles in an ATP and cytoskeleton independent manner reveals a passive, mechanically driven process. This could be a leap forward in the comprehension of the role of caveolae in the cell, and in the understanding of genetic diseases like muscular dystrophies.

Contents

I Introduction   5

1 Physical Description of Cellular Membranes   7
  1.1 Membrane Physics at Equilibrium   7
      1.1.1 Elastic Membrane Properies   8
      1.1.2 Mathematical Description of the Membrane   13
      1.1.3 Membrane Tension   13
  1.2 Techniques to Measure Mechanical Properties of Membranes .   16
      1.2.1 The Micropipette Aspiration Technique   17
      1.2.2 Tether Extraction   20

2 From Vesicles to Cells   29
  2.1 Structure of the Cell   31
  2.2 Cytoskeleton of Cells   33
      2.2.1 Cytoskeleton Filaments   33
      2.2.2 Actin Filaments   34
      2.2.3 Actin Cortex Impairing Drugs   35
  2.3 Cellular Membranes   36
  2.4 Membrane Tension and Endo-/ Exocytosis   37
  2.5 Tether Extraction from Cells   39

3 Caveolae   41
  3.1 The Definition of Caveolae   41
  3.2 The Caveolin Protein   43
      3.2.1 The Structure of Caveolin   44
  3.3 The Cavin Protein Family   46
      3.3.1 Cavin1   47
      3.3.2 Cavin2   48
      3.3.3 Cavin3   48
      3.3.4 Cavin4   48
  3.4 The Assembly of Caveolae   49

	3.4.1	Caveolin is Synthesized in the Endoplasmic Reticulum, and Assembles in The Golgi Aparatus	49
	3.4.2	Cavin Enters the Stage for Caveola Formation	50
	3.4.3	The Lipid Composition of Caveolae	52
	3.4.4	Caveolae are Stable Structures at the Plasma Membrane	53
	3.4.5	Endocytosis of Caveolae	54
3.5	Caveolae/Caveolin Proteins and Signaling Processes		55
	3.5.1	Ion-pumps in Caveolae	55
	3.5.2	eNos in Caveolae	56
3.6	Caveolae in Musclle Cells		56
	3.6.1	Interaction Partners of Cav3 in Myotubes	57
	3.6.2	Muscular Dystrophies	58

4 Mechanical Role of Caveolae   **62**

II Materials and Methods   70

5 Cells and Reagents   **72**
- 5.1 Cell Types and Cell Culture 72
  - 5.1.1 HeLa-PFPIG 73
  - 5.1.2 Mouse Lung Endothelial Cells 73
  - 5.1.3 Mouse Embryonic Fibroblast 74
  - 5.1.4 Human Muscle Cells 74
- 5.2 Treatments Altering the Cell 76
  - 5.2.1 Expression of Proteins 76
  - 5.2.2 Altering Actin Dynamics 77
  - 5.2.3 ATP depletion 77
  - 5.2.4 Cholesterol Depletion 78
- 5.3 Vesicles out of Cellular Plasma Membranes 78
  - 5.3.1 Giant Plasma Membrane Vesicles (GPMV) 79
  - 5.3.2 CytochalasinD - Blebs 80
  - 5.3.3 Plasma Membrane Spheres (PMS) 81

6 Experimental Set-Up   **83**
- 6.1 Tether Extraction 83
  - 6.1.1 Epi-OT 83
  - 6.1.2 Con-OT 84
  - 6.1.3 Cell Stage and Pipette Holder 87
  - 6.1.4 Hypo-osmotic Shock System 88
  - 6.1.5 Fabrication of Micropipettes 89

	6.1.6 Aspiration Controll System	90
	6.1.7 Beads and Bead-coatings	92
	6.1.8 Online Tracking with MatLab	92
	6.1.9 Calibration	94
6.2	TIRF-microscopy	97
	6.2.1 Principle of TIRF	98

III Results   100

7 Tether Extraction From Adherent Cells   102
   7.1 Typical Tether Force Traces 102

8 Preliminary Remarks and Comments on the Relation Between Tether Force and Membrane Tension on Cells   105

9 Caveolae and the Resting Cell Tension   108
   9.1 The Effective Tension of MLEC is Affected by the Presence of Caveolae 108
   9.2 The Effective Tension in MEFs Does not Depend on Presence of Caveolae 110
   9.3 Challenging the Effective Cell Tension by Chemical and Biological Treatments 112
      9.3.1 Alteration of the Cytoskeleton Decrease the Effective Cell Tension 112
      9.3.2 ATP Depletion Decreases the Membrane Tension . . . 113
      9.3.3 Interaction of Cav1 with Src-kinase 113
      9.3.4 Cav3 Re-establishes the Cell Tension of Cav1$^{-/-}$ MLEC 116
   9.4 Summary 118

10 Caveola-mediated Membrane Tension Buffering Upon Acute Mechanical Stress: Experiments on Cells   120
   10.1 Application of Acute Mechanical Stress and Cell REsponse Observed by TIRF and EM 121
      10.1.1 Mechanical Stress Leads to the Partial Disappearance of Caveolae from the Plasma Membrane 121
      10.1.2 Partial Disapperance of Caveolae Observed by EM . . 128
   10.2 Membrane Tension Measurements During Hypo-osmotic Shock 128
      10.2.1 Caveolae are Required for Buffering the Tension Surge Due to Hypo-osmotic Shock 129

  10.2.2 Clathrin Coated Pits do not Buffer the Membrane Tension 131
  10.2.3 Disassembly of Caveolae During Mechanical Stress . . . 135
 10.3 Correlation Between the Observed Loss of Caveolae and the Excess of Membrane Area Required to Buffer Membrane Tension 139

11 Caveola-mediated Membrane Tension Buffering upon Mechanical Stress: Experiments on Plasma Membrane Spheres 142
 11.1 Plasma Membrane Spheres Contain Caveolae and Are Devoid of Actin Filaments 144
  11.1.1 Production of PMS from HeLa-PGFPIG 144
  11.1.2 Production of PMS from MLEC 146
 11.2 Micropipette Aspiration of PMS Induces Disassembly of Caveolae 147
  11.2.1 Quantitative Analysis of Micropipette Aspiration of PMS 150

12 Experiments on Muscle Cells
The Role of Caveolin-3 Mutations in Muscular Dystrophy 156
 12.1 Tether Force of Differentiated Muscle Cells 158
  12.1.1 Reaction of Myotubes with Cav3-Mutations upon Acute Mechanical Stress 158
 12.2 Contracting Myotubes 161

IV Discussion 163

13 Caveolae as a Security Device for the Cell Membrane 164
  13.0.1 Comparison of Experimental Data with the Theoretical Model 167

14 Mechanical Stressand the Role of Caveolae in Signaling 170

15 Towards a Better Understanding of Muscular Dystrophies 172

16 Other Caveolin Related Diseases 175

Part I

Introduction

This work is focused on the mechanical role of caveolae, small plasma membrane invaginations, in cell tension homeostasis. It was pursued in the framework of a close cooperation between the Physics and Biology departments of the Institut Curie, in which biological and physical approaches were combined. Measurements of physical properties of plasma membranes were conducted on living cells, and combined with genetic or chemical cell modifications. This work will be introduced by presenting first the physical background of lipid membranes in model systems, and in the cellular environment. Then, we will summarize the current biological knowledge about caveolae. This part will end with a theoretical outlook on the possible role of caveolae in cell membrane mechanics.

Chapter 1

Physical Description of Cellular Membranes

In this chapter basic physics of pure lipid bilayers will be first described, before moving to systems that are closer to the biological reality (i.e. lipid bilayers with embedded proteins).

1.1 Membrane Physics at Equilibrium

Membranes are quasi two dimensional objects, which can be deformed by:

- shear
- stretch/compression
- bending

In the following we will consider the case of a pure lipid bilayer, which is $\sim 5nm$ thick, and in which a single lipid occupies an area of $\sim 0,7\mu m^2$.

([Lipowsky and Sackmann, 1995]). Lipids consist of a hydrophilic head and a hydrophobic tail. When dissolved in aqueous solution, it is energetically favorable to shield the tail from the water molecules, while the heads prefer to be oriented towards the aqueous solution. This property of the lipids makes them self assemble into bilayers or micelles ([Zimmerberg and Kozlov, 2006]; [Janmey and Kinnunen, 2006]). Theoretical models to describe the abovementioned basic deformations were developed in the 1970's by Helfrich and Canham ([Helfrich, 1973], [Canham, 1970])(Fig. 1.1).

1.1.1 Elastic Membrane Properies

Shear is the deformation of the membrane resulting from the action of two parallel, but opposite directed forces at constant area. The associated density of energy per unit area (H_{shear}) can be deduced from Hook's law:

$$H_{shear} = \frac{1}{2}\mu(\lambda^2 + \lambda^{-2} - 2) \tag{1.1}$$

with the lateral extension rate $\lambda = (L_0 + \Delta L)/L_0$ (L_0 is the starting length of the membrane) and the shear modulus μ (in J/m^2). In the case of fluid lipid bilayers H_{shear} can be neglected due to $\mu = 0$, whereas in the case of biological relevant membranes, where membrane compositions vary locally, and where membrane proteins hinder the flow of lipids, we have to consider $\mu \neq 0$. For example μ was measured to be $6 \cdot 10^{-6} J/m^2$ ([Mohandas and Evans, 1994],[Hénon et al., 1999]).

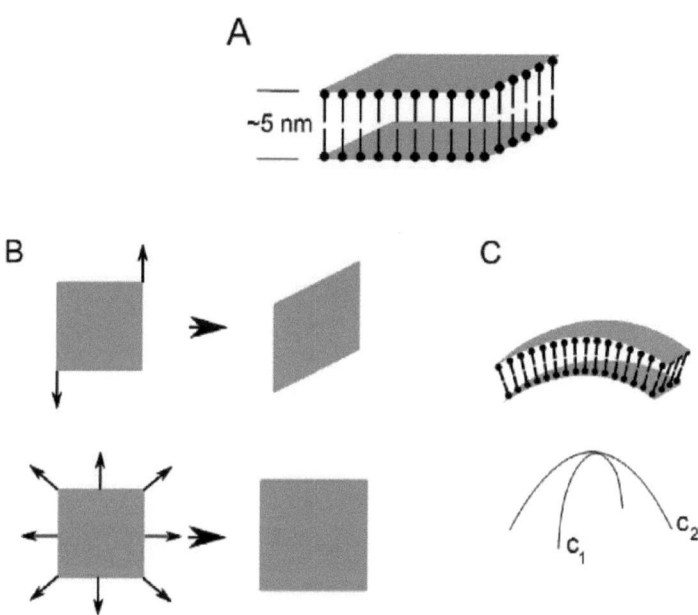

Figure 1.1: Schematic examples of membrane deformations. (A) A lipid bilayer membrane has a typical a thickness of $\sim 5nm$. (B) In plane membrane deformations: shear (top) and stretch (bottom). (C) Bending results in an out of plane deformation and is described by its two principal curvatures C_1 and C_2.

Extension/Compression is the deformation of the membrane with a change of its area of the membrane ΔA. The associated areal density of energy H_{ext} can be expressed as:

$$H_{ext} = \frac{1}{2}K_a(\frac{\Delta A}{A})^2 \quad (1.2)$$

with the elastic area compressibility modulus K_a and the relative area deformation $\Delta A/A$. Lipid bilayers are in general poorly compressible and exensible. K_a is of the order of $200 mN/m$ ([Kwok and Evans, 1981]), and depends weakly on the lipid chain length or the degree of insaturation ([Rawicz et al., 2008]). Increasing the cholesterol concentration of membranes can increase K_a about $2-5$ fold ([Needham and Nunn, 1990]). Upon a few percent of extension, lysis occurs by pore opening. The order of magnitude of the lysis tension is about $10^{-3} N/m$ ([Evans et al., 2003]; [Sandre et al., 1999]).

Bending results in a deformation of the membrane out of its plane, and its corresponding areal energy density can be described as:

$$H_{bend} = \frac{1}{2}\kappa(c_1 + c_2 - c_0)^2 + \kappa_G c_1 c_2 \quad (1.3)$$

with the two principal membrane curvatures $c_1 = \frac{1}{R_1}$, and $c_2 = \frac{1}{R_2}$ representing the shape of the membrane along the two axes at a given point in the membrane, and with the spontaneous curvature c_0 which is considered to be isotropic. κ is the bending rigidity modulus of the membrane, and κ_G is the Gaussian curvature modulus, both expressed in J or $k_B T$. The term $c_1 + c_2$ is called total curvature, the term $c_1 \cdot c_2$ Gaussian curvature. In the case of closed membranes, we can apply the Gauss-Bonnet theorem, which claims, that the the Guassian curvature is an invariant, i.e. its integral does not change until its topology changes (e.g. creation of pores in the membrane). Recently, Baumgart measured the Gaussian bending mudulus ($\kappa_G = 10^{-19} J$) in analyzing the shape of phase seperated vesicles ([Baumgart et al., 2003]). Since the topology of vesicles and cells normally does not change upon our experiments, this term can be omitted.

The bending rigidity modulus, κ, describes the energy required to deform a membrane out of its plane. κ is an intrinsic property of the membrane, ranging typically between $10 - 100 k_B T$. It depends on the membrane composition, as well as on the nature of each lipid like the degree of saturation or the lenght of the alipathic chains ([Evans and Rawicz, 1990a]; [Rawicz et al., 2000a], [Rawicz et al., 2008]). To get an overview of the values of κ in various lipid membranes, and of the different techniques used to measure κ, please see ([Marsh, 2006]). External parameters like the temperature can influence κ, too. Since bending involves the compression and expansion of the two layers of the membrane, κ can be related to the compressibility modulus K_a via:

$$\kappa = K_a h^2 / c_e \qquad (1.4)$$

with the membrane bilayer thickness h and a prefactor c_e, which accounts for the coupling between both monolayers ($c_e = 48$ for uncoupled and $c_e = 12$ for completely coupled monolayers)([Bloom et al., 2009]; [Rawicz et al., 2000a]).

The spontaneous curvature is another intrinsic membrane property, which describes the curvature that the membrane would attain in the absence of external constraints. A membrane made of two identical lipid layers would have no spontaneous curvature. As soon as the two layers are different in their composition or in their distirbution of lipids and proteins, the spontaneous curvature can become non zero, typically of order of a fraction of nm^{-1} ([Kooijman et al., 2005]). The insertion of proteins like epsin or amphiphysin into one leaflet of the bilayer ([Lipowsky et al., 1998]; [Campelo et al., 2008]) or the clustering of lipids induced by the binding of proteins can also induce spoontaneous curvature of order of $(\sim 1/50 nm^{-1})$ ([Römer et al., 2008]).

1.1.2 Mathematical Description of the Membrane

Helfrich and Canham proposed a theoretical description of membranes by introducing a Hamiltonian with membrane bending as the major energy contribution ([Helfrich, 1973], [Canham, 1970]):

$$H = \int dA \frac{1}{2} [\kappa(c_1 + c_2 - c_0^2 + \kappa_G c_1 c_2)] \tag{1.5}$$

However, in order to achieve a more detailed description of the state of the membrane, the tension of the membrane has to be considered, too.

1.1.3 Membrane Tension

The lateral membrane tension, σ, is a parameter of particular interest in the present work. It describes the stress associated to a change of membrane area, $\sigma = \partial F / \partial A$, and can be related to the elastic modulus K_a:

$$\sigma = K_a (\frac{\Delta A}{A}). \tag{1.6}$$

This description is true for the macroscopic scale, but it does not take into account, that the definition of the area becomes delicate in the case of thin membranes under low tension. In our case, the lipid bilayer has a thickness of $h \sim 5nm$ and will be thus subjected to thermal fluctuations on the molecualer and microscopic scale ([Fricke et al., 1986]), which results in a technical impossibility to resolve optically the real membrane area. Thus, membrane area can be stored and released without any visible change resulting in a non linear relationship between applied stress and observed area. One way to resolve this problem is to describe it in analogy to the unfolding of a polymer in solution ([Helfrich and Servuss, 1984]; [Evans and Rawicz, 1990a]). To stretch a Gaussian polymer, one has first to overcome the entropic energy for unfolding before getting into the enthalpic regime, where the extension affects the molecular bonds.

Entropic Regime To find a realtion between the change in area and the membrane tension, one has first to deduce the membrane fluctuations with other membrane properties. Considering a square membrane patch of edge length L under zero tension, one can estimate the maximum amplitude of fluctuations (U_{max}) using the Canham-Helfrich Hamiltonian combined with the equipartition of energy (for more details see ([Helfrich, 1973], [Helfrich and Servuss, 1984])

$$U_{max} \propto \sqrt{\frac{k_B T}{\kappa}} L \qquad (1.7)$$

to be of the order of $3\mu m$ with $\kappa = 10 k_B T$ and $L = 10 \mu m$. Including the membrane tension, one obtains with the same approach ([Helfrich and Servuss, 1984])

$$U_{max} \propto \sqrt{\frac{k_B T}{\sigma} \ln \frac{L}{a}}. \qquad (1.8)$$

with a being a molecular size of the order of $\sim nm$. Regarding the same square membrane patch ($\kappa = 10 k_B T, L = 10 \mu m$) under a typical value of membrane tension ($\sigma = 10^{-6} \frac{N}{m}$) and molecular size ($a = 0,5 nm$) the maximal amplitude of the membrane fluctuations decreases to $0,6\mu m$. Membrane tension thus reduces the amplitude of thermal fluctuations.

More generally, a relationship between the relative memrbane area stored in fluctuations, ($\alpha = \frac{\Delta A}{A}$), and the membrane tension, σ, can be obtained by fluctuation spectrum analysis ([Helfrich and Servuss, 1984]):

$$\alpha = \frac{\Delta A}{A} = \frac{k_B T}{8\pi\kappa} ln \frac{\pi^2/a^2 + \sigma/\kappa}{\pi^2/L^2 + \sigma/\kappa}. \tag{1.9}$$

In the case of zero tension, this equation simplifies to

$$\alpha = \frac{k_B T}{8\pi\kappa} ln \frac{L^2}{a^2}, \tag{1.10}$$

and reveals the excess area to be about 8 % in our example ($\kappa = 10 k_B T, a = 0,5 nm, L = 10 \mu m$). At low membrane tension, i.e. in the range $\kappa\pi^2 \ll \sigma \ll \kappa\pi^2/a^2$, equation (1.1.9) can be simplified to

$$\alpha = \frac{k_B T}{8\pi\kappa} ln \frac{\kappa\pi^2}{\sigma a^2}. \tag{1.11}$$

at $\sigma = 10^{-6} \frac{N}{m}$ the excess area is about 6 %. It is of note that in the entropic regime, the excess area ΔA is the difference between the microscopic (A) and the observed membrane area (A_{obs}), and that the microscopic area remains constant.

Enthalpic Regime

As seen above, membrane tension increase in the entropic regime smoothens out membrane fluctuations. Further increase of σ will lead to a stretch of the lipid bilayer, i.e. the area per lipid molecule will have to increase. The dominant contribution is then gouverned by the elastic modulus and given

by:
$$\frac{\Delta A}{A} = \frac{\sigma}{K_a}. \tag{1.12}$$

The combination of both the contributions from the enthalpic and entropic regimes gives the general equation relating excess area and membrane tension ([Evans and Rawicz, 1990b]):

$$\alpha = \frac{\Delta A}{A} = \frac{k_B T}{8\pi\kappa} ln\frac{\kappa\pi^2}{\sigma a^2} + \frac{\sigma}{K_a}. \tag{1.13}$$

Note that an alternative approach for the definition of the membrane tension was recently proposed by Fournier ([Fournier et al., 2001]). Here the effective membrane tension is associated to the change of optically resolved membrane area, which makes the *ad hoc* introdction of stretching elasticity unneccessary. The herewith obtained relationship between excess area and membrane tension is identical to equation (1.1.13).

1.2 Techniques to Measure Mechanical Properties of Membranes

Various techniques have been developed to assess the mechanical properties of membranes at the microscopic or macroscopic scal. Amongst them, one can cite the vesicle edge detection, where the shape of a fluctuating membrane vesicle is recorded using rapid video microscopy, and the analysis of the fluctuation spectrum reveals the bending rigidity ([Faucon et al., 1989]; [Meleard et al., 1998], [Pécréaux et al., 2004], [Faris et al., 2009]). Another approach analyzes the fluctuations at one point of the vesicle using an optical tweezer ([Betz et al., 2009]). In the following, we will introduce more detailed two other methods the micropipette aspiration, and the tether pulling method that were extensively used in the present work.

1.2.1 The Micropipette Aspiration Technique

In the end of the 1970's, Evans introduced the micropipette aspiration technique on red blood cells ([Waugh and Evans, 1979]) and giant unilamelar vesicles (GUV's)([Kwok and Evans, 1981]) to monitor the membrane area with respect to the applied membrane tension. This enables to determine the bending rigidity κ and the memrbane extension modulus K_a using equation (1.1.13). The basic principle consists in aspirating a fluctuating vesicle in a glass micropipette (typical diameter of $\sim 3 - 5\mu m$). The membrane tension σ, which is tuned by the aspiration pressure, can thus be varied over orders of magnitude. In parallel, the length of the aspirated portion of the vesicle (called "tongue") allows a measure of the excess area α. For proper quantitative analysis, the following requirements have to be met ([Drury and Dembo, 1999]):

- the voume of the vesicle remains constant

- the vesicle contains a constant number of lipids

- the membrane does not adhere to the pipette

- the aspirated tongue is longer than the pipette radius R_{pip}

More precisely, the membrane tension imposed by the micropipette aspiration to the vesicle can be derived using the Laplace law:

$$\sigma = \frac{\Delta P R_{pip}}{2(1 - R_{pip}/R_{ves})} \qquad (1.14)$$

with the hydrostatic pressure difference ΔP between the inside and the outside of the pipette, and R_{pip} and R_{ves} the radii of the pipette and the vesicle, respectively. in practice, the hydrostatic pressure is set by varying the height of a water reservoir connected to the micropipette (see materials and methods section for more details). The change of excess membrane area, i.e. the change of the length of the tongue $\Delta L = L - L_0$ can be linked to the change of total membrane area before (A_0) and after (A_{asp}) aspiration using geometrical arguments ([Kwok and Evans, 1981]):

$$\frac{A_{asp} - A_0}{A_0} = \frac{(R_{pip}/R_{ves})^2 - (R_{pip}/R_{ves})^3}{2R_{pip}} \Delta L. \qquad (1.15)$$

Reminding that the relative excess area stored in fluctuations is $\alpha = \dfrac{A - A_{obs}}{A}$, with A the microscopic area and A_{obs} the observed area, this value will be maximal (α_0) in the case of a non aspirated vesicle (σ_0). Increasing the micropipette aspiration, and thus the applied membrane tension $\sigma > \sigma_0$ will lead to a decrease of the relative excess area $\alpha < \alpha_0$ following equation (1.1.11). The relation $(\alpha_0 - \alpha) = \dfrac{A_{asp} - A_0}{A_0}$ allows linking the measurable parameters R_{pip}, R_{ves} and ΔL to the membrane tension, and to the bending rigidity and the extension moduli ([Evans and Rawicz, 1990a], [Rawicz et al., 2000a], [Fournier et al., 2001]):

$$\alpha_0 - \alpha = \frac{k_B T}{8\pi\kappa} ln\frac{\sigma}{\sigma_0} + \frac{\sigma}{K_a} \qquad (1.16)$$

At low tension the logarithmic term dominates, giving acces to κ through a linear fit of the $ln\sigma$ vs. $(\alpha_0 - \alpha)$ plot. At higher tension, the linear term will dominate, which allows to deduce K_a via a linear fit of the σ vs. $(\alpha_0 - \alpha)$ plot. This technique has been widely used, especially by the Evans' group, to measure mechanical properties of synthetic and biological membranes ([Mohandas and Evans, 1994], [Hochmuth, 2000], [Manneville et al., 2001], [Herant et al., 2005], [Tian et al., 2007], [Rawicz et al., 2008]).

1.2.2 Tether Extraction

Another way to achieve informations about mechanical membrane properties consists in extracting tethers by the application of a "point" force on the membrane. Tether were first observed in the early 1970's on red blood cells attached on a glass slide and submitted to shear flows ([Hochmuth et al., 1973]). Variations of this approach invlolved the introsuction of micropipettes to controll the membrane tension ([Hochmuth et al., 1982]), the study of artifical membranes ([Waugh, 1982], [Rossier et al., 2003]) or the use fo gravity instead of shear flow ([Bo and Waugh, 1989]). Tethers have also been formed b using of kinesin motors walking on microtubules ([Roux et al., 2002], [Koster et al., 2003]), beads attached to the sensitive force probes suc as the biomembrane force probe ([Heinrich et al., 2005]) or magnetic tweezers ([Heinrich and Waugh, 1996]) or optical tweezers ([Dai et al., 1998], [Koster et al., 2003], [Cuvelier, 2005]). In the present work, we have used optical tweezers: optically trapped beads serve as handles to extract tethers from adherent cells as well as from membrane vesicles aspirated by a micropipette.

Force and Radius of a Tether

In this section, the exp erimental situation of a tether extracted from an aspirated vesicle (1.2.2) will be discussed, because this case has been extensively investigated both from a theoretical and experimental point of view ([Svetina et al., 1998]; [Evans and Yeung, 1994]; [Derényi et al., 2002]; [Powers et al., 2002]; [Rossier et al., 2003]).

The free energy (F_t) of a tether (i.e. a cylindrical membrane tube) can

be written:
$$F_t = \frac{\pi}{R_t}\kappa L_t + 2\pi r\sigma l L_t - fL_t \qquad (1.17)$$

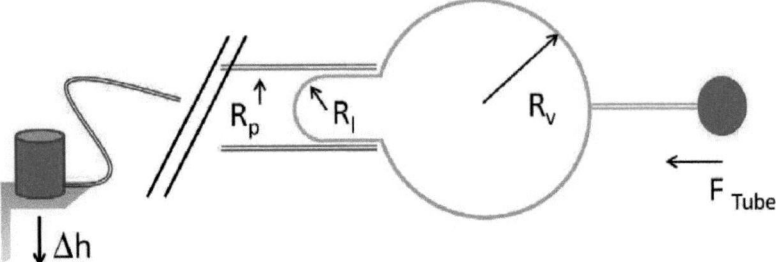

Figure 1.2: Scheme of an aspirated vesicle with radius R_v from which a tether of radius R_t and length L_t is extracted under the application of the tether force f_t.

with R_t the tether radius, L_t the tether length, and f the force required to hold the tether at constant length. This term reflects the competition between the bending rigidity leading term, which favors small curvature, and the membrane tension leading term tending to minimize the tether area. At equilibrium, at constant tether length, the tether force f_0 and the tether radius R_0 are obtained by minimizing the free energy with respect to R_t and to L_t:

$$f_0 = 2\pi\sqrt{2\kappa\sigma} \tag{1.18}$$

$$R_0 = \sqrt{\frac{\kappa}{2\sigma}}. \tag{1.19}$$

One may remark that f_0 and R_0 only depend on the bending rigidity and the membrane tension, but not on the tether length L_t providing σ is fixed. Equation 1.2.6 was experimentally checkked by Heinrich and Waugh using magnetic tweezers ([Heinrich and Waugh, 1996]). For typical values for the bending rigidity, $\kappa = 10 k_B T$, and the membrane tension, $\sigma = 10^{-6} \frac{N}{m}$, the equilibrium force is about $10pN$ and the radius about $140nm$. If the membrane tension does not remain constant or is not set by micropipette aspiration during tether extraction, the tether force becomes length dependent according to ([Cuvelier, 2005]):

$$L \simeq R_v \frac{f_t R_v}{16\pi\kappa}[\frac{k_B T}{\kappa}ln\frac{f_t}{f_0} + \frac{f_t^2 + f_0^2}{\pi\kappa K_a}] \qquad (1.20)$$

Tether Nucleation

Tether pulling is mainly c omp osed of two phases: nucleation and elongation. The nucleation process has been studied theoretically by Derenyi and colleagues ((([Derényi et al., 2002]) and exp erimentally by Koster et al. ([Koster et al., 2005]). First, the flat membrane is deformed to a catenoïd shape, which is the minimal surface between two spheres. At this stage, the force to elongate the catenoïd depends linearly to its length. At longer deformations, nucleation into a cylindrical membrane tube (i.e. tether) becomes energetically favorable and the tether elongation becomes independant from its length. The transition between the flat membrane and the tether is accompanied by a force overshoot, whose amplitude depends on the size of the adhesion patch of the membrane to the bead.

In the case of living cells, the nucleation process also involves membrane from cytoskeleton detachment. Depending on the tether pulling velocity, tether force traces may or may not exhibit force overshoots ([Heinrich et al., 2005]). In the present work, only tethers will be considered which were preformed and elongated further.

Dynamics of Tether Extraction

Equation 1.2.6 can be refined by considering additional contributions applicable in specific situations:

- asymmetry between the two leaflets:
  When a long tube is pulled, the difference between the outer and inner leaflet is not negligible. For a tube with the length of $100\mu m$ and a bilayer thickness of $5nm$ the difference is in the order of $3\mu m^2$. The supplementary contribution to the force is

  $$f_{nlb} = \frac{2\pi \kappa_r}{e}\alpha,$$

  where α is the difference of membrane area between the two leaflets, and κ_r the nonlocal curvature ([Bozic et al., 1992]; [Raphael and Waugh, 1996]). Experimentally, $k_r \approx 3\kappa$ ([Svetina et al., 1998]), and Δf_{nlb} is thus in the range of $1pN$, and still small compared to the typically observed tether forces $(10 - 50pN)$. Moreover, it is a transient effect, because the flip-flop process of lipids will equilibrate the number of lipids within minutes ([Svetina et al., 1998]).

- viscous friction:
  Experimentally tethers are not extracted infinitely slowly. The viscous friction between the two leaflets of the membrane,, which arises upon extraction at finite velocity v, has to be considered. Since the flow of lipids into the tether is equal for the two monolayers, the lipids in the inner leaflet have a different distance to go than the ones in the outer

leaflet. Evans and Yeung ([Evans and Yeung, 1994]) have shown that the additional force f_v is expressed as

$$f_v = \eta_m e^2 \ln(\frac{R}{r_t}) v_{pull},$$

where η_m is the viscosity between the two leaflets, R is the radius of the vesicle and v_{pull} is the velocity of tube elongation. For $v_{pull} = 0.5 \mu m/s$ (as used in our experiments) and $\eta_m = 0.002 \frac{pNs}{\mu m}$ ([Hochmuth et al., 1996]), one obtains $\Delta f_v \sim 0.03 pN$, which is negligible. Note however that these estimates are only valid for pure lipid bilayers.

Chapter 2

From Vesicles to Cells

The equations and techniques presented in the previous section were experimentally validated on synthetic lipid membrane systems, especially on giant unilaminar vesicles (GUV's). These systems are especially suited to study physical parameters, and to test theoretical assumptions, since the lipid composition of the membrane and their geometry. Additionally, due to the size of GUV's (diameter $\sim 30 - 30\mu m$), they easily can be manipulated and subjected to mechanical forces in a controlled and tunable manner.

In contrast to the model systems, the cellular plasma membrane is a complex dynamic system with a large diversity of lipids and proteins organized as an asymetric bilayer and constantly subjected to renewal. In the present work, we will exploit the knowledge obtained on model systems to explore the mechanical properties of the cellular plasma membrane probed by the using the tether pulling technique based on optical tweezers. We will first describe briefly the physicist's view of a cell. In particular, we will focus on the plasma membrane and the differences that we can anticipate as compared with pure lipid bilayers.

2.1 Structure of the Cell

The basic building unit of living organisms is the cell. Although organisms consist of a single cell type (i.e. embryonic stem cells) at the very beginning, these cells are able to divide and differentiate into a large variety of cell types with different shapes, and specializations for different tasks. However, the general organizing mechanisms, and the machinery needed for proper cell function are similar in the different cell types. Eukaryotic cells (i.e. cells with a nucleus like plant, fungi and animal cells) are typically of the size of $5 - 500 \mu m$. Interestingly, even at these small scales, there is a clear organization of spatially and functionally separated intracellular compartments (organelles) to address different cellular functions. In these compartments, different components and functions are segregated from each other: DNA storage in the nucleus, ATP production in the mitochondria, synthesis of proteins and lipids in the endoplasmic reticulum, and sorting of proteins for their final destinations in the Golgi apparatus.

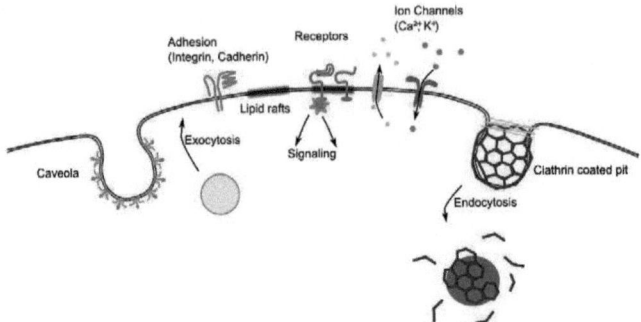

Figure 2.1: Sketch of the cellular plasma membrane with regions whose organization can be divided in three categories. (1) The various lipids in the plasma membrane form regions of different composition (so called lipid rafts), (2) single proteins, either prefering lipid rafts or not, act as signalling platforms, ion channels or adhesion sites, and (3) supramolecular structures like the clathrin coated pits or caveolae are composed of muliple proteins and a specific lipids composition and play a role in endocytosis or signaling platforms. Renewal of the plasma membrane is established by fusion of vesicles coming from the inside of the cell (exocytosis), and removal of the plasma membrane occures through budding and pinching off of the plasma membrane into te cell (endocytosis).

Cells, and the organelles within them, are separated from each other and from the outside world by thin membranes. The main building block of these cellular membranes are lipids and proteins (Fig 2.1). The plasma membrane lis on another important cellular component, the cytoskeleton.

2.2 Cytoskeleton of Cells

2.2.1 Cytoskeleton Filaments

The main functions of cytoskeleton are to maintain the cell shape and to drive cell motility (in using structures like pseudopods or flagella). It also plays an important role in transport processes (intracellular transport, endocytosis, and exocytosis) and in cell division. It is a dynamic structure containing three main kinds of polymer filaments (in eukaryotic cells): microtubules, intermediate filaments and actin filaments ([Fletcher and Mullins, 2010]).

Microtubules are hollow cylinders of about $25 nm$ diameter, commonly built up by 13 protofilaments, which are polymers of α and β tubulin. They bind GTP for polymerization and show a very dynamic behavior. Microtubules are organized in the centrosome and represent the stiffest cytoskeletal component (persistence length $\sim 1mm$)([Bretscher, 1991]).

Intermediate filaments have a diameter of $8 - 11nm$ and are less rigid than microtubules. They organize the internal three dimensional structure of the cell (e.g. the nuclear envelope) and can be made of like vimentin ([Strelkov et al., 2003]).

Actin filaments have a persistence length of $\sim 13.5 \mu m$, a diameter of $7nm$, and are made of two actin chains organized in a helicoidal structure. Actin filaments are highly concentrated beneath the cell membrane. This cortex, which interacts directly with the cell membrane, controls the cell shape and some endocytic processes.

2.2.2 Actin Filaments

Dynamics of Polymerization and Depolymerization

Actin filaments are polarized. Due to the arrowhead pattern created when myosin binds actin, the rapidly growing end of a filament is called the barbed end, the slowly growing one the pointed end. The polymerization of actin requires hydrolysis from ATP to ADP (i.e. energy consumption), and is regulated by the chemical equilibrium between the concentration of actin monomers and actin filaments. The process of continuous addition of actin monomers to the barbed end of the filament and its depolymerization at the pointed end is called treadmilling. *In vitro* studies of the concentration of unpolymerized actin found in cells ($10-100\mu M$), and determined elongation rates of barbed ends of $\sim 0.3 - 3\mu m/s$, that is $\sim 100 - 1000 subunits/s$. To control this process, cells use two mechanisms including proteins that bind monomeric actin and modify its polymerization properties (e.g. profilin, thymosin-β 4)([Chuat et al., 1996]), and capping proteins that prevent the polymerization from the monomers by adding to the filament ends (e.g. gelsolin). Furthermore, several proteins promote actin nucleation (e.g. Arp 2/3) by forming complexes with membrane proteins (e.g. the signaling pathway through WASp/Scar to Arp 2/3 complex)([Fletcher and Mullins, 2010]; [Pollard and Cooper, 2009]; [Pollard et al., 2000]).

Besides the WASP-pathway, there are other interactors connecting the actin filaments to the cell membrane, e.g. the phosphatidylinositol 4,5-biphosphate (PIP2), filamin or ezrin. The elasticity of the actin cytoskeleton is due to both the cross linking of actin filaments involving proteins called cross linkers like α-actinin or ezrin, and by the work of molecular motors, especially myosin 2, enabling a contraction of the cytoskeleton ([Paluch et al., 2006]).

2.2.3 Actin Cortex Impairing Drugs

Several drugs are known to interfere with actin treadmilling by distinct mechanisms.

- **Cytochalasin D** caps the barbed end of actin filaments preventing both assembly and disassembly of subunits. Owing to the dynamic nature and to additional severing of actin filaments, cytochalsin D treatment leads ultimately to actin filament disassembly([Fujimoto et al., 2000]; [Franki et al., 1992]; [Cooper, 1987]; [Schliwa, 1982]).

- **Latrunculin A/B** forms complexes with actin monomers and thereby shifts the chemical equilibrium between actin monomers and filaments, which also results in depolymerization of actin filaments ([Wakatsuki et al., 2001]).

- **Jasplakinolide** stabilizes actin filaments in retaining phosphatic groups from the ATP hydrolysis in the filament. Since the addition of further actin monomers on the barbed end is not affected, jasplakinolide induces actin filament polymerization until the pool of actin monomers is depleted ([Holzinger and Meindl, 1997]; [Senderowicz et al., 1995]; [Bubb et al., 1994]).

Note that the drug effects described above are valid for the concentrations applied in the experiments described further below $(C \sim 0, 1 - 1\mu M)$. Some subtle eddects of these drugs might arise at other concentrations.

2.3 Cellular Membranes

Cellular membranes consist of many different lipids and these membranes typically contain embedded (membrane) proteins and associated proteins ([Janmey and Kinnunen, 2006]). Cell membranes are very thin ($\sim 5nm$), and are therefore flexible (i.e. easily deformed by forces in the pN regime).

Membrane proteins either interact with each other or with specialized lipids ([Zimmerberg and Kozlov, 2006]). In addition, proteins that are solubale in the cytosol (i.e. the intracellular fluid) can interact directly with the membrane lipid domains or proteins. A crucial property of lipid bilayers is their two-dimensional liquid nature (above a certain temperature). This allows for the lateral diffusion of molecules in the membrane, which is essential for proper cell functioning. Even though the bilayer behaves in many respects as a two-dimensional fluid, the molecular mobility within the membrane can be restricted by several factors. First, the lipid bilayer of cellular plasma membranes has an asymmetric lipid and protein composition, whic is actively maintained through transport of lipids by proteins (flippases) from one layer to the other ([Buton et al., 1996]). Second, membranes may contain substructures often referred to as rafts ([Hancock, 2006], [Simons and Toomre, 2000]). Depending on the lipid composition of the membranes, phase separation between liquid ordered and liquid disordered phase domains may be induced. Third, the underlying cortical cytoskeleton and associated proteins limit the diffusion of molecules in the membrane by creating some obstacles (fences) (Akihiro2005).

2.4 Membrane Tension and Endo-/ Exocytosis

The role of the cytoskeleton in determining the viscoelasticity of a cell and in governing cell motility, growth and developent has been extensively studied in the past ([Vogel and Sheetz, 2006]; [Park et al., 2005]; [Geiger et al., 2001]; [Stamenovic and Wang, 2000]). While the cell membrane was often considered to be a passive border between the interior of the cell and the outside world, it has recently become evident that the plasma membrane plays also an active role in various cellular processes ([Morris and Homann, 2001]).

In this context, important parameters are the membrane surface area and the lateral membrane tension which are interconnected and related to processes like the release (exocytosis) or uptake (endocytosis) of cargo, to cell migration or to the repair of the cell membrane. Sheetz and coworkers showed in red blood cells, that a hypo-osmotic shock- hence a stretch of the membrane surface area- increased the membrane tension and resulted in a decrease of endocytosis. Contrary, stimulation of exocytosis dropped the membrane tension, which was later (after 5 min of stimulation) compensated by increased endocytosis ([Dai et al., 1997]). A similar observation was also done on cells under mitosis, where endocytosis was impaired at high membrane tension during mitosis, but started again after the mitosis when membrane area increased and membrane tension was low ([Raucher and Sheetz, 1999]). Thus, reduction of membrane area increases membrane tension and impairs endocytosis, and exocytosis increases the membrane area and lowers the membrane tension. Another mean to control the membrane tension and to resist mechanical stresses consists in the cytoskeleton underneath the plasma membrane ([Sheetz et al., 2006]; [Mills and Morris, 1998]). In moluscan neuron cells, mechanosensitive channels are more sensitive to membrane stretch after disruption of the underlying cytoskeleton. At the other hand, force mediated f-actin recruitment at the plasma membrane of fibroblasts increases the membrane rigidity ([Glogauer et al., 1997]). A third way to manage rapid increases of membrane area consists in the release of excess membrane area stored in folds or invaginations at the plasma membrane. Electrophysiologic measurements of the capacitance of whole cells revealed that erythro cytes can undergo swelling up on hypo-osmotic shock in smoothing their membrane. Mast cells and lymphocytes were reported to double their volume at constant capacitance ([Solsona et al., 1998]; [Ross et al., 1994]). In skeletal muscle cells almost half of the membrane area is stored in caveolae (with a density of $\sim 40 \mu m^2$), which may ensure the enormous large membrane stretches of muscle cells in action by flattening ([Dulhunty and Franzini-Armstrong, 1975]). The subject of the present work is the role of caveolae as a membrane reservoir, and how this is related to the membrane tension.

2.5 Tether Extraction from Cells

The theory of tether pulling presented in the previous chapter was developed and experimentally validated for lipid vesicles. The complexity of the cellular plasma membrane and its interaction with the cytoskeleton make it necessary to examine how this formalism is changed for living cells. Sheetz and colleagues have studied the effect of the interaction between plasma membrane and cytoskeleton on the membrane tension, especially the role of PIP_2 (phosphatidylinositol 4,5-bisphosphate), a small GTPase in the cell membrane, as a regulator of cytoskeleton-cell membrane adhesion ([Sheetz et al., 2006]; [Raucher et al., 2000]). Similar studies were also conducted on cells treated with diverse drugs affecting the actin cytoskeleton, the microtubules or the membrane lipid composition ([Raucher and Sheetz, 1999]; [Hochmuth et al., 1996]). To explain the tether force measurements on cells, an additional term, which accounts for the cytoskeleton-cell membrane adhesion, is introduced into the free energy of the tether, i.e.

$$F_t = \frac{\pi}{r}\kappa l + 2\pi r l \sigma + 2\pi r l W_0 - fl \qquad (2.1)$$

This leads to an equilibrium force given by:

$$f^* = 2\pi\sqrt{2(\sigma + W_0)\kappa} = 2\pi\sqrt{2\sigma_{eff}\kappa} \qquad (2.2)$$

where W_0 is the cytoskeleton-membrane adhesion energy density (J/m^2). It can be considered for tether pulling experiments on cells in normal conditions that W_0 contributes to about 75 % to the effective tether force and

σ to the remaining 25 % ([Sheetz, 2001]). It is worth noting that the adhesion of the cytoskeleton to the membrane is described by a single continuum parameter, W_0, and that more detailed models of the underlying molecular processes are still missing. For example protein-lipid and protein-protein interactions within the membrane may lead to compartmentalization and selfassembly of peculiar structures like membrane invaginations, which are likely to alter the properties of the cell membrane ([Zimmerberg and Kozlov, 2006]; [Hinrichsen et al., 2006]). The mechanical role of these structures is poorly understood, but there is growing evidence that cells control the membrane tension and contain some membrane reservoirs to allow fast stretching without membrane rupture (e.g. in muscle cells). Since these cells contain a large amount of membrane invaginations called caveolae, P. Sens proposed a theoretical model how those invaginations could play the role of a membrane reservoir and influence the plasma membrane mechanical properties ([Sens and Turner, 2005]). The theoretical model will be described in section 4 (page 74). But first will be presented some basic biological knowledge about caveolae.

Chapter 3

Caveolae

3.1 The Definition of Caveolae

Caveolae are small Ω - shaped invaginations of $50 - 80 nm$ diameter. They were described the first time in the 1950's, when Palade and Yamada observed small, cave-like membrane invaginations of mouse cells with electron microscopy ([Palade, 1953]; [Yamada, 1955]). In the 1990s, the identification of the caveolin protein family necessary for caveola formation boosted the insights into the molecular structure of caveolae, which are additionally rich in cholesterols, glycosphingolipids and sphingomyelins ([Monier et al., 1995]). Even more recently, additional proteins were discovered, called the cavin-protein family, whose interaction with caveolins is important for the formation and shape of budded caveolae ([Hansen and Nichols, 2010]). An example of caveolae and of its composition is shown in Fig 3.1.

Although caveolae are known since more than 50 years, their biological function is still under debate. They are proposed to play a role in signal-

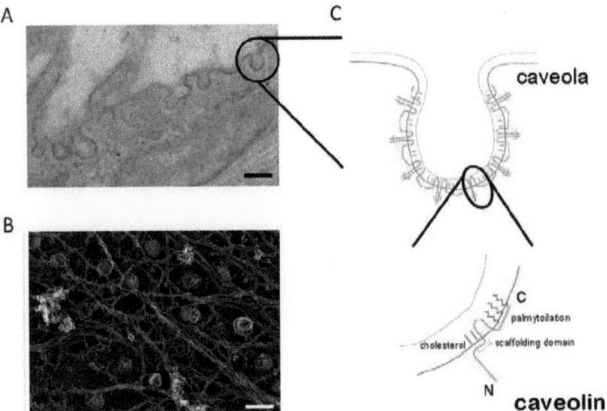

Figure 3.1: (A) Electron micrograph of an ultrathin cryosection obtained from a mouse lung endothelial cell showing invaginations typical for caveolae. Bar = 200nm. by G. Raposo (B) Deep etch electron microscopy image showing cell membrane caveolae from the inside of mouse lung endothelial cells. Bar = 200nm. by M. Morone (C) The sketch indicates how caveolin is inserted into the caveolar membrane, with the N and C termini facing the cytoplasm and a "hairpin" domain embedded within the membrane bilayer. The scaffolding domain, a highly conserved region of caveolin, might have a role in cholesterol interactions. The C-terminal domain, which is close to the intramembrane domain, is modified by palmitoyl groups that insert into the lipid bilayer ([Parton et al., 2006]).

ing, transcytosis, endocytosis, cell migration, mechanical sensing, lipid and cholesterol transport, tumor suppression or promotion, depending on the studied cell type ([Goetz et al., 2008]; [Parton and Simons, 2007]). In contrast to this, the focus of the present work was to experimentally check the hypothesis that caveolae act as a membrane reservoir and constitute a mean of membrane tension buffering ([Sens and Turner, 2005]; [Dulhunty and Franzini-Armstrong, 1975]). In the following, a short description will be given how caveolin and cavin proteins were identified and what their characteristics are. An updated description of the caveolin-proteins can be found in ([Goetz et al., 2008] and [Parton and Simons, 2007]), and for the cavin-proteins read ([Nabi, 2009] and [Hansen and Nichols, 2010]).

3.2 The Caveolin Protein

The first protein which was found to be necessary for caveola formation is called caveolin, and was discovered in the early 1990s by several research groups. Rothberg and coworkers found an antibody which labeled a 22kDa protein and associated to caveolae proved by ultra structural immunogold electron microscopy. Rapid freeze techniques were used to observe a flattening of caveolae after treatment of cells with fillipin and nystatin accompanied by the dissociation of the 22kDa protein. Due to the intimate association of this protein to caveolae it was called caveolin ([Rothberg et al., 1992]). In an attempt to study the vesicle transport and sorting between the apical and basolateral membrane of epithelial cells, Simons and coworkers cloned a protein called VIP-21 (vesicular integral protein of 21kDa), which was found in the membrane of the trans Golgi-vesicles ([Kurzchalia et al., 1992]). It turned out, that this protein had the identical sequence as the caveolin protein found by Rothberg, thereby showing, that it is at the same time a structural component of caveolae, and plays a role in the vesicular traffic. In the following further iso-forms of caveolin were found, which were grouped in the caveolin protein family: the ubiquitously expressed caveolin-1 (former VIP-21) and caveolin-2 ([Scherer et al., 1996]) as well as the muscle specific caveolin-3 ([Way and Parton, 1995]). In the following caveolin proteins will be called **cav1, cav2** and **cav3**.

3.2.1 The Structure of Caveolin

Cav1 is a 178 amino acid protein, which is expressed in most cell types, but especially in endothelial, epithelial, cardiac muscle cells and adipocytes ([Vogel et al., 1998]; [Thorn et al., 2003]). Cav2 contains 161 amino acids and is co-expressed with cav1 ([Scherer et al., 1996]). Cav3 is made of 151 amino acids and is specific for muscle cells. Additionally to the full length proteins (α isoform), both cav1 and cav2 have shorter sized isoform. The β isoform of cav1 lacks the first 32 amino acids, the β and γ isoforms of cav2 are less well studied. A detailed compare of the amino acid sequence of the three caveolin-family members is shown in [Razani et al., 2002].

As mentioned above, the cav1 protein is made of 178 aminoacids, and its single regions together with their interaction with other proteins or lipids will be described in the following. Even if the studies were done with the full length cav1, the most remarks stay true for the other members of the caveolin protein family.

The caveolin protein has a amphipathic N- and C-terminus and a central part, which is buried into the lipid bilayer of the Golgi or the plasma membrane. The end of the N-terminus is marked by a syronine S80 in cav1 followed by a scaffolding domain (CSD) of about 20 amino acids at residues 82 to 101 (in cav2 54-73, and 55-74 in cav3), which facilitates interaction with other caveolin proteins and numerous other proteins, such as *src* family tyrosine kinases, growth factor receptors, G-protein-coupled receptors (GPCRs), G proteins and endothelial nitric oxide synthase (eNOS) ([Quest et al., 2008]; [Head and Insel, 2007]). At the other hand, it is considered that the CSD of caevolins forming a caveola are close to or buried into the plasma membrane, and thus cannot interact anymore with other proteins. A part of the CSD contains a potential cholesterol recognition and/or interaction amino acid consensus (CRAC) motif (residues 94-101, [Epand et al., 2005]). This motif is not implicated in a direct 1:1 interaction with cholesterol, but it is considered to improve the binding of cav1 to the plasma membrane and dense cholesterol packing in caveolae. Residues 101-126 are amphiphilic and thus buried into the lipid bilayer of the plasma membrane. Three cysteins are on the C-terminus at residues 133, 143, and 156. The palmytoilation of these sites is not necessary for membrane binding of caveolin, but is considered to stabilize it ([Parton et al., 2006]).

3.3 The Cavin Protein Family

First, the caveolin proteins were considered as the one and only necessary to form caveolae, but in 2008 and 2009, the cavin protein family was identified to be essential for the formation and shape of caveolae. The work of Vinten and colleagues in 2001 showed some evidence for a protein associated to caveolae ([Vinten et al., 2001]), but it was Parton and coworkers to show that a protein formerly known as polymerase I and transcript release factor (PTRF) is necessary for successful formation of budded caveolae at the plasma membrane. Further they reported that in the absence of cavin1 budded caveolae are not detectable via EM, and that the amount of diffusing cav1 proteins in the plasma membrane increases ([Hill et al., 2008]). Short time after, three other proteins we re found to be implied in shaping caveolae ([Liu and Pilch, 2008], [Aboulaich et al., 2004], [McMahon et al., 2009], [Bastiani et al., 2009]). As a result of these findings, the four proteins with their newly discovered role in caveola formation were regrouped in the cavin protein family. All cavins have numerous phosphorylation sites, giving rise to interaction with PKC, and bind to immobilized phosphatidylserine. Due to this, cavins also appear often with multiple bands in western blots, which made it until now impossible to measure their molecular weight, but estimated values are 55kDa for cavin1, 72kDa for cavin2, 43kDa for cavin3, and 43kDa for cavin4. They share a PEST (proline, glutamic acid, serine, and threonine-rich) domains, and also have a leukine zipper domain (aa 50 - 98 in cavin1, aa 52 -100 in cavin2, and aa 20 - 78 in cavin3) through which they might bind to the cav1 or cav3 protein ([Hansen et al., 2009]). It is interesting that the cavins were not identified earlier as caveola related proteins, which might have been due to their weak binding to the caveolar membrane, and hence the technical difficulty to detect the interaction.

3.3.1 Cavin1

Cavin1 was first identified in a yeast two-hybrid screen, where it was shown to interact in the nucleus with both the transcription-termination factor-I and the polymerase-I, and to enable polymerase-I transcription. That's why it was initially called polymerase I and transcription release factor (PTRF)([Grummt, 1999]; [Leary and Huang, 2001]). Whether there is a link between the caveola associated localization of cavin1 and its role in the nucleus, will have to be studied in the future. The first hint for an interaction of cavin1 and caveolae was given by Vinten and colleagues who identified cavin1 as a caveola protein and observed a correlation between the expression of cav1, cavin1 and the abundance of caveolae. ([Vinten et al., 2005]; [Voldstedlund et al., 2001]). They also showed that cavin1 interacts with both cav1 and cav3, and that its distribution at the plasma membrane is broader than the one of caveolins. Hill and coworkers showed using FRET (fluorescence resonance energy transfer) between fluorescently tagged cavin1 and cav1 molecules that they have to be in close contact to form budded caveolae, and that in the absence of cavin1 the fraction of free diffusing cav1 in the plasma membrane increases, whereas the cellular amount of cav1 decreases ([Hill et al., 2008]).

3.3.2 Cavin2

Cavin2 was separately identified as a phosphatidylserine (PS)-binding protein from human platelets ([Burgener et al., 1990]) and as a serum deprivation response (SDPR) ([Gustincich and Schneider, 1993]). A first report of association of cavin2 to caveolae was given in 2000 ([Rybin et al., 2000]). The group of Nichols showed in 2009 that cavin2 is involved in the formation of caveolae, and that it can induce tubulation of the plasma membrane when over expressed ([Hansen et al., 2009]).

3.3.3 Cavin3

Cavin3 was described as a Sdr-related gene product that binds to C-kinase (SRBC), and as a phosphatidylserine-binding protein that is induced during serum starvation ([Izumi et al., 1997]). The association of cavin3 to cavin1 and their relation with caveolae was indicated in 2004 by Aboulaich and coworkers using vectorial proteomics and indirect fluorescence microscopy ([Aboulaich et al., 2004]).

3.3.4 Cavin4

In contrast to the three first members of the cavin family, cavin4 is considered to be muscle specific and was first reported as a muscle-restricted, coiled-coil protein (MURC) ([Ogata et al., 2008]). Although that work reported already a similar localization of cavin4 in striated muscle as the muscle specific cav3, an explicit connection between these proteins was not shown until 2009 ([Bastiani et al., 2009]).

3.4 The Assembly of Caveolae

3.4.1 Caveolin is Synthesized in the Endoplasmic Reticulum, and Assembles in The Golgi Aparatus

Caveolin is synthesized in the endoplasmic reticulum (ER), and oligomerizes in the ER to form SDS-resistant homo-oligomers of 7-14 caveolin proteins ([Hayer et al., 2010]; [Fernandez et al., 2002]). Monier and colleagues concluded from their experiments that caveolin oligomerizes to complexes of 14-16 proteins in the ER prior to completion of Golgi transit ([Monier et al., 1995]). At the other hand, there is evidence that free caveolin proteins exist in the Golgi. Bush and colleagues analyzed the labeling of caveolin by different antibodies, and reported that some antibodies only recognize the Golgi-pool of caveolin, whereas others recognize the plasma membrane proteins assembled in caveolae. Since the antibody recognizing the CSD labeled the Golgi-pool, it was suggested that caveolin oligomerizes after the Golgi, which renders the CSD inaccessible for the antibody ([Bush et al., 2006], [Pol et al., 2005]). Pelkmans and Zerial, and more recently Helenius and coworkers conclude from their experiments that caveolin oligomerizes during its way through the Golgi, assembles cholesterol and glycosphingolipids at the Golgi membrane to form exocytic carrier vesicles. These vesicles are transported to the plasma membrane, where they fuse with it to form caveolae. GFP tagged cav1 was used to analyze quantitatively the fluorescent signal of quantal caveolar structures at the plasma membrane, and the intensity of such a single quantum was estimated to correspond to 144 ± 39 cav1-GFP molecules ([Bauer and Pelkmans, 2006]; [Pelkmans and Zerial, 2005]). At the stage of oligmerization, caveolins can either form homo-oligomers of cav1 or hetero-oligomers of cav1 and cav2. As mentioned above, cav2 alone cannot form caveolae, and it needs the association to cav1 to be transported to the plasma membrane ([Song et al., 1997]).

3.4.2 Cavin Enters the Stage for Caveola Formation

At which state of the caveola formation the cavins become important is not very clear yet, but there is some evidence that cavin1 interacts with cav1 and cav3 at or short after the Golgi-level, whereas cavin2 and cavin3 have a localization more restricted to the plasma membrane ([Hayer et al., 2010]). A reduction of cavin1 results not only in a reduction of budded caveolae, but also decreases the total amount of cav1 and cav3 proteins in the whole cell. The size of caveolin agglomerates shrinks, whereas the fraction of free caveolin proteins increases ([Hill et al., 2008], [Liu and Pilch, 2008]). Since cavin1 affects not only cav1 but also the muscle specific cav3, animals or humans with a defect in cavin1 show the phenotype of a cav1/cav3 double knock-out ([Liu et al., 2008], [Hayashi et al., 2009]). The over expression of cavin1 causes a higher number of budded caveolae without any change in the expression of cav1 and cav3, respectively. These findings suggest that cavin1 acts as a regulator for the cell to tune the number of budded caveolae at the plasma membrane.

A decrease in cavin2 expression results also to a reduction of budded caveolae due to both, a decrease of the cav1 level as well as of the level of cavin1. This makes it more difficult to isolate the effect of cavin2 on caveola formation, but it seems that cavin2 is responsible for the shape of caveolae, whereas cavin1 is there to bud them in. An overexpression of cavin2 does not increase the number of caveolae, but alters their appearance from a Ω-shape to a more elongated structure up to tubule like structures with the caveolin proteins at the tip. Another interesting finding shows that cavin2 has to interact with cavin1 to bind to the caveola membrane ([Hansen et al., 2009]). Cavin3 is the third partner of this complex involved in caveola formation, but its role is even less clear. It co-precipitates together with cavin1 and cavin2, and needs either cav1 or cav3 for binding at the plasma membrane. In contrast to cavin1 and cavin2, cavin3 binds also to other caveolin containing structures such as caveosomes, and seems to play a role in the traffcking of caveolin inside the cell. According to this, it was observed that intracellular cav1 traffic is dramatically reduced in the absence of cavin3 ([Hansen et al., 2009]).

Finally, the musc le specific cavin4 localizes together with cav3 and interacts with cavin1 and cavin3 ([Bastiani et al., 2009]). Over expression of cavin4 results in contractile dysfunction in heart muscle, and studies with a reduced expression of cavin4 are not yet published. At the Golgi, none of the cavins localizes with caveolins, whereas it was shown by FLIM/FRET and co-immunoprecipitation that cavins form a multimeric complex in the cytosol which interacts at the plasma membrane with caveolin oligomers to form a caveola ([Bastiani et al., 2009], [Hansen and Nichols, 2010]). For a summary see also [Nabi, 2009].

3.4.3 The Lipid Composition of Caveolae

The lipid composition of caveolae is not very well studied, and the results may depend on the way, how plasma membrane caveolae were separated from the rest of the cellular membranes. Pike and colleagues studied the composition of detergent resistant (considered as raft like) and nonresistant membranes of human epidermal carcinoma cells without caveolae and cells transfected with cav1 and containing caveolae. They found an increase of cholesterol in the detergent resistant membranes of cells with caveolae, i.e. $\sim 600 \pm 200 nmol/mg\, protein$ of cholesterol in caveola/raft membranes compared to $sim 500 \pm 100 nmol/mg\, protein$ of cholesterol in raft membrane without caveolae . Interestingly, there was no significant change in the lipid composition ([Pike et al., 2002]). Örtegren and colleagues analyzed the lipid composition of adipocyte caveolae, and report that each $100 nm^2$ of caveolar membrane contain 192 cholesterol molecules, 65 sphingmyolin lipids, 206 glycerophospholipids, and 6 protein molecules. Compared with the noncaveolar membrane, caveolae are highly enriched in cholesterol and have about twice the amount of sphingomyolin (see Fig. 3.4.3)([Örtegren et al., 2004]). Thus a caveola of $60 nm$ diameter formed by 144 caveolin proteins contains about 22000 cholesterol molecules, i.e. a cholesterol to caveolin ratio of about 157 : 1!

Estimated number of lipid and protein molecules per 100 nm² of plasma membrane bilayer inside and outside caveolae.			Calculation for a caveola of 60 nm diameter

Component	Caveolae	Plasma membrane outside caveolae	
Cholesterol	192	70	~22000
Sphingomyelin	65	35	~7500
Glycerophospholipids	206	208	~23000
Protein	4	5	~460

Figure 3.2: Lipid composition of caveolar and non-caveolar membrane of adipocytes. from [Örtegren et al., 2004]

3.4.4 Caveolae are Stable Structures at the Plasma Membrane

Once a caveola is formed, it stays as a stable structure as it does not exchange single caveolin proteins with its neighbor caveolae or cytosolic caveolins. This stability was studied by several groups with FRAP (fluorescence recovery after photobleaching) experiments and analysis of caveola dynamics with TIRF. Thomsen and colleagues conducted the first FRAP experiments on cells expressing GFP labeled cav1. After bleaching regions with several caveolae a very slow recovery of the caveolin signal was observed (after 10 minutes, fluorescence intensity recovered to $\sim$ 30%) ([Thomsen et al., 2002]). This led to the conclusion, that there is no exchange of single caveolin proteins between a caveola and the cytosolic pool. Tagawa and colleagues studied this stability more precisely in bleaching single caveolae formed by cav1-GFP, and observed even after 15 minutes only little recovery of the fluorescence signal.

In a second approach, they fused two cells, one expressing cav1-GFP and the other expressing cav1-RFP, which allowed them to follow the dynamics of caveolin proteins between single caveolae. Already formed caveolae stayed single colored even after long time (3 hours)([Tagawa et al., 2005]). These experiments affirmed the idea, that the caveolin proteins are tightly held together in a caveola, and do not exchange with their environment. Concerning the mobility of the caveolae, three populations of equal size were observed. One staying stable on the plasma membrane over several minutes, one moving laterally on the plasma membrane and one shortly connecting and detaching to the plasma membrane (so called kiss and run movement)([Pelkmans and Zerial, 2005]). Alterations of the cytoskeleton dynamics or altering the amount of cholesterol in the plasma membrane can influence the caveolar dynamics.

3.4.5 Endocytosis of Caveolae

One of the proposed roles of caveolae is the endocytosis, i.e. uptake of cargo into the cell. Several studies identifid amongst others albumin, cholera toxin B (Ctx-B), and simian virus 40 (SV-40) as c argoes using caveolae to enter the cell, but none of them is specific for caveolae, i.e. they can also use other pathways to enter the cell ([Botos et al., 2007], [Nabi and Le, 2003], [Pelkmans et al., 2001]). Caveola endocytosis depends on the phosphorylation of tyrosine residues of proteins associated to the caveola and implies the activation of *src*-kinases. Thereafter, the surrounding actin filaments are disassembled to reorganize as an actin patch at the caveola, followed by bursts of actin polymerization forming transient actin "tails" (of $1.5\mu m$ length). Depending on the cargo, dynamin is recruited to the neck of the loaded caveola and pinches it off through hydrolyzation of GTP to GDP ([Sverdlov et al., 2007]). Additionally, cell polarization and migration seems to depend on the interaction of integrins with cav1 and in the regulation of caveolar endocytosis ([Grande-García and del Pozo, 2008], [Echarri et al., 2007]).

3.5 Caveolae/Caveolin Proteins and Signaling Processes

Many signaling proteins, and channels are reported to be enriched or clustered in caveolae, which led to the picture that caveolae act as a signaling platform. Additionally, cav1 is interacting through its caveolin scaffolding domain (aa 82-101) with a variety of signaling proteins including *src*-family, protein (PKA, PKB, PKC) and other kinases, the insulin receptor, eNOS, H-Ras, G proteins and G protein coupled receptors (GPCR)([Patel et al., 2008]) gives an overview of the supposed signaling processes which involve caveolae. A few examples will be presented in the following.

3.5.1 Ion-pumps in Caveolae

Fujimoto showed 1993 in an electron microscopy study, that Ca^{2+}-pumps are $18 - 25$ fold enriched in caveolar structures compared to the other plasma membrane in several mouse cell types and in human fibroblasts ([Fujimoto, 1993]). Other studies report a localization of some K^+-, and Na^+-channels in caveolae ([Kristensen et al., 2008]; [Maguy et al., 2006]; [Daniel and Cho, 2006]; [Bergdahl and Swärd, 2004]; [Taggart, 2001]).

3.5.2 eNos in Caveolae

One of the most studied binding partners with caveolin is the endothelial nitric oxide synthase (eNOS). The CSD interacts with eNOS and inhibits its enzyme activity, and knock-out of cav1 upregulates eNOS expression. Caveolae concentrate cellular eNOS, which is thought to allow the cell to inhibit eNOS activity under resting conditions, and to allow rapid activation upon mechanical stimulation ([Balligand et al., 2009]).

3.6 Caveolae in Musclle Cells

Caveolae are very abundant in muscle fibers and store about half of the plasma membrane area ([Dulhunty and Franzini-Armstrong, 1975]). Briefly, the formation of muscle fibers starts with the differentiation and fusion of myoblasts to form myotubes. Myotubes are contractile, and bundles of them constitute a muscle fiber ([Cossu et al., 1996]). The expression pattern of many proteins changes like the myogenic factors Myf5 or MyoD during the differentiation of myoblasts and fusion to myotubes ([Yamamoto et al., 2008]). It is for note that in the case of skeletal and cardiac muscle cells the expression of cav1 and cav2 switches to cav3 during differentiation ([Song et al., 1996]). Additionally, the muscle specific cavin4 is expressed and associates with the other cavin proteins at caveolae, but little is known about its role in caveola formation ([Bastiani et al., 2009], [Ogata et al., 2008]). During contraction the plasma membrane of muscle cells is injured by shearing stress and stretch, and several strategies exist in muscles to avoid or to repair those injuries ([Brown and Glover, 2007]). Besides the seal of breaches by muscle precursor cells ([Cornelison, 2008]) myotube internal strategies ensure membrane integrity. With respect to the scope of this work, the dystrophin-glycoprotein complex, dysferlin, and cav3 are presented as well as related muscular dystrophies.

3.6.1 Interaction Partners of Cav3 in Myotubes

Besides the formation of caveolae, cav3 is found to be part of the dystrophin-glycoprotein complex (DGC)([Song et al., 1996]; [Blake et al., 2002]) and to colocalize with dysferlin at the plasma membrane ([Lennon et al., 2003];[Hernández-Deviez et al., 2006]).

The Dystrophin-glycoprotein Complex

The role of the DGC is to link the extracellular matrix with the cytoskeleton and to provide structural stability to the plasma membrane ([Durbeej and Campbell, 2002]; [Dalkilic and Kunkel, 2003]). It is formed by dystrophin, sarcoglycans, dystroglycans, sarcospan, syntrophin, dystrobrevin and cav3 ([Blake et al., 2002]). β-dystroglycan is the only component of the DGC which directly interacts with dystrophin. Cav3 binds to the same motif of β-dystroglycan, and competes with dystrophin. The DGC is unstable in the absence of dystrophin and cells become abnormally susceptible to damage from contraction ([Danialou et al., 2001]; [Petrof et al., 1993])(Fig 3.6.1).

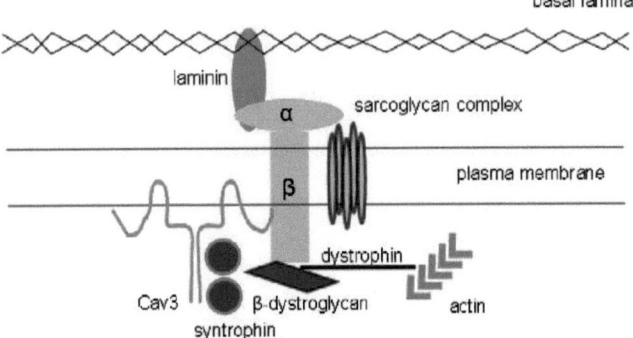

Figure 3.3: Sketch of the dystrophin-glycoprotein complex and its associates proteins at the myotube plasma membrane. Dystrophin and cav3 compete in binding to β-dystroglycan ([Galbiati et al., 2001]).

Dysferlin

Dysferlin is important for rapid membrane repair of muscle cells. The repair depends on Ca^{2+} and ensures the addition of intracellular vesicles on a sub-second to second timescale ([Bansal et al., 2003], [Lennon et al., 2003]). Dysferlin interacts with annexins, calpain-3, and affixin, which are involved in actin cytoskeleton remodel ing ([Brown and Glover, 2007]), but needs cav3 for the transport to and correct organization at the plasma membrane ([Hernández-Deviez et al., 2006]).

3.6.2 Muscular Dystrophies

Muscular dystrophies are genetically caused muscular disorders characterized by progressive weakness and loss of integrity. Symptoms like high serum levels of creatine kinase can occur from any age from birth to midlife ([Laval and

Bushby, 2004]). More than 30 gene loci related to muscular dystrophies were identified reflecting the molecular diversity underlying muscle formation. In the following the Duchenne Muscular Dystrophy (DMD) and dystrophies related to cav3-mutations will be described.

Duchenne Muscular Dystrophy

Duchenne muscular dystrophy (DMD) is one of the most severe myopathies. Patients are wheelchair bound before 12 years old and die in their third decade of life. DMD is caused by a lack of dystrophin, which causes the instability of the DGC and, thus, a higher fragility of the plasma membrane of myotubes. The lack of dystrophin leads to a downregulation of the DGC associated sarcoglycans and dystroglycans ([Sicinski et al., 1989]). In contrast, cav3 expression is 2 − 3 fold upregulated ([Vaghy et al., 1998]), and electron microscopy studies showed an increase of plasma membrane caveolae ([Bonilla et al., 1981]). The reason for this upregulation of cav3 is not known.

Cav3-mutations and Related Muscular Dystrophies

Cav3 is important for the formation of myotubes, associates with the DGC, forms caveolae and is necessary for proper dysferlin localization at the plasma membrane. At the other hand are mutations of cav3 associated to diverse forms of muscular dystrophies such as limb girdle muscular dystrophy (LMGD), rippling muscle disease (RMD), hyper-CKemia (HCK), long QT syndrome and sudden infant death syndrome ([Gazzerro et al., 2010]). The three best studied diseases- LGMD, RMD and HCK- are described more detailed in the following.

LGMD is a heterogeneous group of muscular disorders which comprises six autosomal dominant (type 1) and nine autosomal recessive (type 2) genes ([Bushby, 1999]). LGMD is defined by a weakness and wasting in the pelvic and shoulder muscles starting in the second or third decade of life and with relatively slow progression. Several mutations of cav3 were shown to cause the autosomal dominant LGMD-1 C ([Galbiati et al., 2001]), of which four were analyzed more carefully and reported to cause retention of cav3 in the Golgi and an impaired targeting to the plasma membrane: (1) cav3-P104L is a missense mutation within the membrane-spanning domain ([Carozzi et al., 2002]), (2) cav3- ΔTFT lacks three amino acids within the caveolin scaffolding domain ([Sotgia et al., 2003]), and the missense mutations in the N-terminal domain (3) cav3-R26Q ([Sotgia et al., 2003]), (4) cav3-A 45T ([Herrmann et al., 2000]).

RMD is a rare autosomal muscular disorder, which manifests by mechanical hyper-sensitivity, i.e. mechanical stimuli induce contractions of electrically silent muscles. Additional symptoms are muscle stiffness and exercise-induced cramps. Interestingly, amongst the cav3 mutations found in patients were the point mutations R26Q, A45T and P104L, which were also reported in cases of LMGD-1C ([Betz et al., 2001]).

HCK describes elevated levels of serum creatine kinase, which is a general sign for myopathies. Creatine kinase is a cytosolic muscle enzyme, which is released into the blood after lysis or necrosis of muscle fibers. However, HCK can occasionally be found in healthy individuals. Patients with idiopathic HCK do not show any muscle weakness or other abnormalities on neurologic examination, electromyography or muscle biopsy ([Gazzerro et al., 2010]). The point-mutation cav3-R26Q, which is also associated with LGMD-1C and RMD, is associated with HCK as well as cav3-P28L, which is a less severe mutation, and leads to a 65% reduction of cav3 expression compared to wt cells ([Merlini et al., 2002]). It is for note that the increasing numbers of diseases related to mutations of caveolin are grouped as "caveolinopathies". In summary, the reported muscular dystrophies associated to cav3-mutations are less severe than the DMD. Why and how the muscular dystrophies occur is not yet understood on a mechanical or molecular basis. According to the role of the DGC to stabilize the plasma membrane, it is suggested that DMD is due to a higher injury rate, which overwhelms the repair capacities of the muscle, and thus leads to loss of it. The observed increase of caveolae could be a reaction of the cell to compensate the increased stress of the plasma membrane. In the case of cav3-related mutations it is not clear, if the muscular dystrophies and HCK are due to an increased rupture of the plasma membrane caused by the lack of a membrane reservoir formed by caveolae. Another reason could be the lack of the repair mechanism established by dysferlin. The study of the short time response of myotubes to acute mechanical stress performed in this work aims to clarify this issue.

Chapter 4

Mechanical Role of Caveolae

P. Sens and M. Turner ([Sens and Turner, 2005]) have recently proposed that caveolae could act as a membrane reservoir, and thus play a role in cell tension regulation. The core of the Sens and Turner model consists in investigating the response of lipid domains that tend to be invaginated under low membrane tension, and flat under high membrane tension. We reproduce here the key points of their reasoning, as presented in ([Sens and Turner, 2005]). For simplicity, it is assumed that the membrane tension σ changes linearly with its area variations. Extraction of membrane area A_T (e.g. by pulling a tether or stretching the cell membrane) increases the membrane tension, and thus the rate of domain flattening, which in turn releases membrane area, and decreases membrane tension again. The variation of membrane tension with the extracted tether area A_T reads:

$$\sigma = \sigma_0 + K_m(A_T + A_{res} - A_{res}^{(0)}) \qquad (4.1)$$

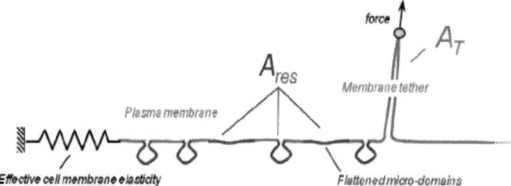

Figure 4.1: Sketch depicting the mechanical response of a membrane to the pulling of a tether. Pulling an area A_T out of the cell either increases the membrane tension σ (the length of the spring), or triggers membrane exchange with a reservoir (?attening of invaginated membrane domains)([Sens and Turner, 2005])

with σ_0 the membrane tension without a tether, and K_m the stretching modulus of the membrane. A_{res} (without tether $A_{res}^{(0)}$) is the membrane reservoir which is stored in the invaginated membrane domains (Fig 4).

The energy of a domain has three contributions arising from line tension, bending and surface tension effects:

$$f(\beta) = 2\sqrt{\pi S}\gamma\sqrt{1-\beta} + 8\pi\kappa\beta + \sigma S\beta \qquad (4.2)$$

$$= f_{\gamma,\kappa}(\beta) + \sigma S\beta, \qquad (4.3)$$

where κ is the bending rigidity of the domain membrane, disfavoring the budded state, is the line tension on the neck of the domain, promoting invagination, and S is the area of a domain. The parameter β characterizes the domain shape ($\beta = 1$ for a fully budded domain, and $\beta = 0$ for a flat domain). At vanishing membrane tension ($\sigma = 0$), the budded state becomes energetically favorable above a critical domain area $S_c = \pi(\frac{4\kappa}{\gamma})^2$.

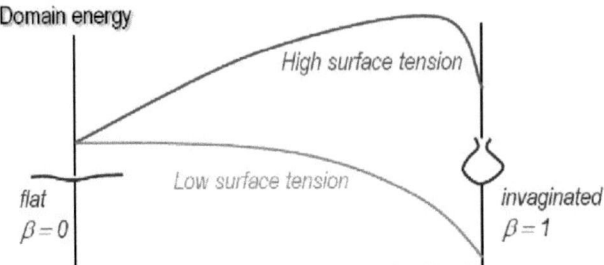

Figure 4.2: Energy of a domain for low and high membrane tension. Small tensions favor budded shape ($\beta = 1$), and large tensions favor ?at shape ($\beta = 0$) of the domain. Above a certain membrane tension, the budded shape remains a local minimum, and an energy barrier has to be overcome to reach the flat shape ([Sens and Turner, 2005]

For biological typical values ($\kappa = 20 k_B T$, $\gamma = 1 \dfrac{k_S T}{nm}$) S_c is about $0.02 \mu m^2$, which is similar to the size of a caveola (corresponding to a sphere with a diameter of $80nm$). With increasing membrane tension, the energy of a budded domain ($\beta > 0$) increases, and above a critical membrane tension, β^*, the flat state becomes stable:

$$\sigma^* S = f_{\gamma,\kappa} \qquad (4.4)$$

$$(\beta = 0) = -f_{\gamma,\kappa} \qquad (4.5)$$

$$(\beta = 1) = 2\sqrt{\pi S}\gamma - 8\pi\kappa \qquad (4.6)$$

It is worth mentioning that, at higher membrane tensions, the budded and flat domain shapes are separated by an energy barrier. This allows a coexistence between the two states of the domains, and is necessary for their role as membrane tension regulators ((Fig. 4.2). Taking $S = 0.1\mu m^2$ the energy scale of the system are

$$\bar{\gamma} = \sqrt{\pi S} \qquad (4.7)$$

$$\gamma \simeq \bar{\kappa} = 8\pi\kappa \simeq 500 k_B T \quad \bar{\sigma} = \sigma S \qquad (4.8)$$

The flat state becomes stable for $\bar{\sigma} = 500 k_B T$ corresponding to $\sigma = 2\dot{1}0^{-5}\dfrac{J}{m^2}$ (consistent to measured cell membrane tensions ([Sheetz, 2001]), and since the energy barrier between flat and budded state is big compared to the thermal energy $k_B T$, the transition between the two states is discontinuous, and the domains snap open rather than continuously flatten up on increasing membrane tension. Considering an ensemble of N domains, the membrane reservoir area reads

$$A_{res} = NS\epsilon\beta_{bud} \qquad (4.9)$$

with the fraction of budded domains ϵ. The total energy of the system is

$$\begin{aligned} F &= N\{\{\epsilon f(\beta_{bud}) + (1-\epsilon)f(0)\} + \int dA\gamma(A) & (4.10) \\ &= N\{\epsilon[\sqrt{\pi S}\gamma\sqrt{1-\beta_{bud}} + 8\pi\kappa\beta_{bud} + \sigma S\beta_{bud}] + (1-\epsilon)[\sqrt{\pi S}\gamma]\} + \int dA (4.11) \end{aligned}$$

Optimizing the energy for the fraction of budded domains $\partial F/\partial \epsilon = 0$ leads directly to the regulation of membrane tension to the value σ^*, which depends on the characteristics of the membrane reservoir (γ, κ and S), but not on the extracted membrane area:

$$\partial F/\partial \epsilon = N\{\sqrt{\pi S}\gamma\sqrt{1-\beta_{bud}} + 8\pi\kappa\beta_{bud} + \sigma S - \sqrt{\pi S}\gamma\} = 0 \quad (4.12)$$

with $\beta_{bud} = 1$ expression(4.12) leads to the same membrane tension as σ^* in (4.6),

$$\sigma^* S = 2\sqrt{\pi S}\gamma - 8\pi\kappa \quad (4.13)$$

To obtain the kinetic response of the system, the transition of the domains between budded and flat state is described as a classical Kramer's process, where the transition time depends exponentially on the energy barrier ΔE that has to be overcome in the process:

$$\tau = \tau_0 exp(\Delta E/k_B), \quad (4.14)$$

where τ_0 is the characteristic time of flattening (here assumed to be the same for both processes).

In a tether pulling experiment, where the extraction is performed at constant velocity ($v_{pull} = 0.5\mu m/s$), we consider the reservoir response to the membrane area extraction rate $\dot{A}_T = 2\pi r_{tether} v_{pull} = 0.08 \mu m^2/s$ (tether radius $r_{tether} = 50nm$). The force of such a dynamical perturbation is influenced by the viscous dissipation (e.g. around the cytoskeleton anchors), and the kinetic response of the reservoir. Here, we only account for the latter effect. In the case of a slowly applied perturbation (i.e. $\dot{A}_T \tau_0 \ll A_{res}$), which resembles the experimental situation, the reservoir has time to equilibrate.

Considering a constant membrane tension ($\frac{\partial \bar{\sigma}}{\partial t} = 0$) due to regulation during tether extraction, an analytical expression of the plateau height can be obtained

$$\bar{\sigma}_{1\to 0} = \bar{\sigma}^0 - (1-\beta_{bud})^{-3/2}\sqrt{2\bar{\sigma}}k_B T \ln(\frac{A_{res}}{2\dot{A}_T \tau_0}), \qquad (4.15)$$

where $\beta_{bud} < 1$ characterizes the finite neck size, and with the flattening membrane tension

$$\bar{\sigma}^{(0)} = S\sigma^{(0)} = (1-\beta_{bud})^{-1/2}\bar{\gamma} - \bar{\kappa} \qquad (4.16)$$

Thermal fluctuations smooth the transition between budded and flat state with the consequence that the tension is not perfectly constant during the transition, and the slope of mid plateau is of the order

$$\frac{\partial \sigma^\star}{\partial A_{T|Plateau}} \simeq \frac{4k_B T}{S A_{res}} \qquad (4.17)$$

But in our case, for typical values $S = 0.1\mu m^2$ and $A_{res} = 30\mu m^2$ this effect is negligible. This theoretical framework suggests that caveolae act as a membrane reservoir, and may explain how they influence the membrane tension of cells. To test this hypothesis, extraction of tethers and simultaneous monitoring of the tether force were performed on cells at rest and after application of acute mechanical stress (e.g. hypo-osmotic shock). In another approach, membrane vesicles containing caveolae were created, and micropipette aspiration combined with monitoring of the membrane tension were carried out to study more structural information about the membrane reservoir formed by caveolae:

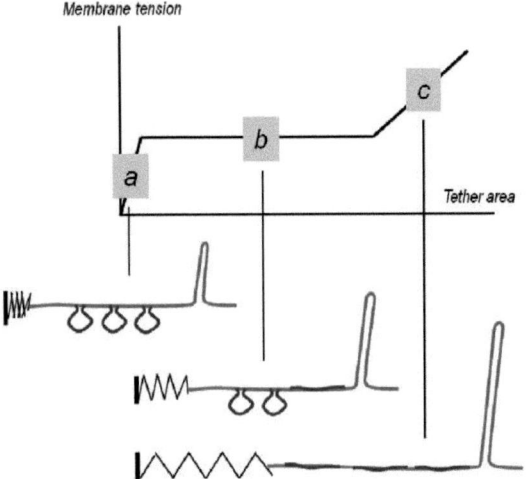

Figure 4.3: By Pulling a tether from a membrane with caveolae, the surface tension increases (a) until it reaches the critical value. Further pulling then opens the caveolae, and releases additional membrane, which allows the extraction of further membrane at constant surface tension (b). If the reservoir is emptied, membrane pulling increases the membrane tension (c)

- the height of the force plateau is related to the energy required to flatten a domain, and depends on the properties of the domains (γ, κ and S)

- the length of the plateau corresponds to the size of the accessible membrane reservoir that is the amount of budded domains.

Part II

Materials and Methods

The main technique used in this thesis is the optical tweezers-assisted tether pulling assay. This method was occasionally combined with other techniques like micromanipulation with micropipettes and/or confocal or epi fluorescence microscopy. Additionally, cells were imaged by total internal reflection fluorescence (TIRF) microscopy, and cell biology techniques were used extensively like addition of reagents altering specific cellular properties or cell transfection to induce the expression of specific proteins. With the aim to create an artificial model system, protein purification and reconstitution in lipid vesicles were tried, and specific cell treatments leading to the formation of vesicles made from native membranes were performed. The detailed protocols are delegated to the Appendix.

Chapter 5

Cells and Reagents

5.1 Cell Types and Cell Culture

Experiments were done on four different cell types:

- HeLa-PFPIG

- mouse lung endothelial cells (MLEC)

- mouse embryonic fibroblast (MEF)

- human muscle cells

The cells were cultured and treated (transfection, plating on microscope slides...) in the biology laboratory. At least 24h before tether extraction experiments, cells were transferred to the physics laborattory and deposited in the cell incubator.

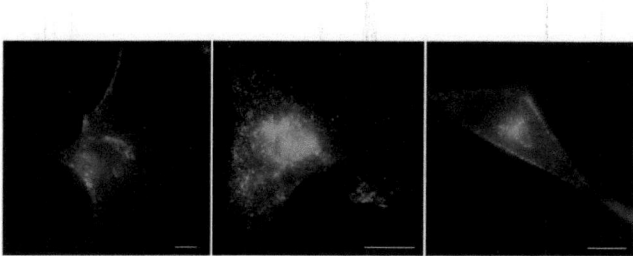

Figure 5.1: Immunofluorescence of fixed cells labeled with cav1-anitbody. Starting from left: wt-MLEC, wt-MEF, HeLa-PFPIG. Bar = $10\mu m$.

5.1.1 HeLa-PFPIG

HeLa-PFPIG cells were developped, characterized and kindly provided by F. Pinaud ([Pinaud et al., 2009]). They stand out with their stable expression of EGFP (enhanced green fluorescent protein) tagged cav1 allowing fluorescence imaging studies (see Fig 5.1).

Cells were maintained in DMEM supplemented with 10% FBS (Invitrogen, Carlsbad, CA) + $100U/ml$ penicillin + $100mg/ml$ streptomycin + $2mM$ glutamine at $37°C$ in $5\%CO_2$.

5.1.2 Mouse Lung Endothelial Cells

Wildtype (wt) and cav-1 negative (cav1$^{-/-}$) mouse lung endothelial cells (MLEC) were characterised by Sessa and coworkers ([Murata et al., 2007]), and were kindly given by R. Stan (Dartmouth Medical School, NH, USA). Wt cells stand out with a very high caveola density (see Fi 5.1, 5.1.3), whereas the cav1$^{-/-}$ cells are devoid of caveolae. Both cell lines were very useful to gain direct inside in the influence of the presence or absence of caveolae on the cellular response.

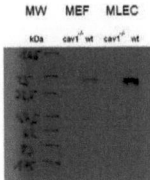

Figure 5.2: Immunoblotting labeling Cav1 equal amounts of cell lysates: wt / Cav1$^{-/-}$ MEF and wt / Cav1$^{-/-}$ MLEC.

Cells were maintained in EGM-2 (Lonza, Basel, Switzerland) supplemented with 20% fetal bovine serum (Thermo Fisher Scientific, Waltham, MA) + $100U/ml$ penicillin + $100mg/ml$ streptomycin at $37°C$ in $5\%CO_2$.

5.1.3 Mouse Embryonic Fibroblast

Wt and cav1$^{-/-}$ mouse embryonic fibroblasts (MEF) were developped and characterised by several laboratories (for example [Razani and Lisanti, 2001]) and were kindly provided by T. Kurzchalia (Max-Planck-Institue of Molecular Cell Biology and Genetics, Germany). Wt cells express cav1 but at a lower level than wt MLEC (see Fig 5.1, 5.1.3.

Cells were maintained in DMEM supplemented with 10% FBS (Invitrogen, Carlsbad, CA) + $100U/ml$ penicillin + $100mg/ml$ streptomycin + $2mM$ glutamine at $37°C$ in $5\%CO_2$.

5.1.4 Human Muscle Cells

Human muscle cells were all prepared and provided by A. Bigot in the group of G. Buttler-Browne (Institut de Myology, Paris, France). While the wt cell line as immortalized, the other cell lines were obtained from biopsies of patients with cav3 mutations and muscular dystrophies (the biological material came from the Muscle Tissue Culture Collection, Munich, Germany):

- Cav3-R26Q: associated with LGMD [code: Bochum 62/01; male *1987; mutation: not communicated]

- Cav3-P28L: associated with hyper C Kemia [code: Bonn 18/03; female *1969; mutation heterozygot 27C>A (27 D>E)]

- Cav3-A45T: associated with RMD [code: Bochum 1/98; male *1936; mutation: not communicated]

For proliferation, cells were maintained in X medium (64% DMEM + 16% 199 Medium + 20% FBS + $2,5 ng/ml$ HGF (all Invitrogen, Carlsbad, CA) + $50 \mu g/ml$ gentamycin + $10^{-7} M$ Dexamethasone (Sigma-Aldrich, St. Louis, MO). Note that muscle cells are first in a undifferentiated state. These so called myoblasts still express cav1 and cav2, but no cav3. Differentiation can be induced, during whichh 2 - 10 cells fuse to form myotubes, and shift to cav3 expression ([Halayko et al., 1999]; [Yamamoto et al., 2008]). For that, cells were densly plated on collagen covered glass slides and incubated for > 7 days at $37°C$ in $5\% CO_2$ in differentiation medium, i.e. DMEM (Invitrogen, Carlsbad, CA) + $50 \mu g/ml$ gentamycin + $10 \mu g/ml$ of bovin Insulin + $100 \mu g/ml$ of human Apotransferrin (all Sigma-Aldrich, St. Louis, MO).

5.2 Treatments Altering the Cell

5.2.1 Expression of Proteins

MLEC were transiently transfected with DNA plasmid, and siRNA, respectively, using AMAXA HUVEC nucleofector kit (Lonza, VPB-1002) following the manufacturer's instructions, and used for experiments 24-72 hours post transfection. Briefly, for each transfection 10^6 detached cells were suspended in $100 \mu l$ of Amaxa-nucleofector reagent supplemented with $0,5-3\mu g$ DNA plasmid, filled into a special cuvette which was placed into the transfection machine, and transfection was induced using program $A-34$. Afterwards, the cells in solution were put into a cell culture flask (T-25) containing $10ml$ prewarmed medium and kept at $37°$ and $5\%CO_2$.

Muscle cells were transfected with DNA plasmid using Fugene6 (Roche, Indianapolis, USA) using a protocol adapted from ([Malik et al., 2000]). Briefly, cells were grown to $\sim 80\%$ confluence on collagen coated glass slides ($2*10^5$ cells per $25mm$ diameter slide). For each transfection a mix of $9\mu l$ Fugene6 with $3\mu l$ DNA plasmid in a final volume of $1ml$ serum free medium was prepared and incubated for $15min$ at room temperature. Then, cells were rinsed with fresh serum free medium, gently covered with the Fugene-DNA mix, and centrifugated for $30min$ at $1180g$ and $32°C$. Afterwards, cells were left in the Fugene6-DNA mix for $4h$ at $37°C$ and $5\%CO_2$ before the transfection solution was replaced by fresh differentiation medium. Cells were differentiated, and were ready for experiments $7-10$ days after transfection.

DNA Plasmids and SiRNA used during this study:

- Cav1-EGFP ([Pelkmans et al., 2001])

- Cav1-Y14F-EGFP ([Orlichenko et al., 2006])

- Cav3-GFP ([Hill et al., 2008])

- Cav3-P104L-YFP ([Couchoux et al., 2007])

- Cavin-mCherry ([Hill et al., 2008])

- Lifeact-mCherry ([Riedl et al., 2008])

- clathrin-heavy-chain siRNA (CLTC-siRNA, Thermofisher)

5.2.2 Altering Actin Dynamics

In order to alter the actin filament structure and dynamics, cells were incubated for 5-20 min in their culture medium supplemented with 0,1% BSA and either $1\mu M$ Cytochalasin D (CD; Sigma-Aldrich, St. Louis, MO)([Schliwa, 1982]), $1\mu M$ Latrunculin A (Lat A; Sigma-Aldrich)([Rotsch and Radmacher, 2000]), and $1\mu M$ Jasplakinolide (Jas; Molecular Probes, Invitrogen, OR, USA) ([Senderowicz et al., 1995]), respectively.

5.2.3 ATP depletion

ATP depletion was performed by washing the cells with PBS^{++} (Phosphate Buffer Saline + $1,5mM\ Ca^{++}$ + $1,5mM\ Mg^{++}$) followed by incubation for 30 min at $37°C$ in PBS^{++} supplemented with $10mM$ deoxy-D-glucose and $10mM NaN_3$ ([Römer et al., 2008]).

5.2.4 Cholesterol Depletion

For cholesterol depletion of the plasma membrane cells were incubated for $50 min$ in PBS^{++} supplemented with $5 mM$ m-β-cyclodextrin (MβCD) (Sigma-Aldrich) which resulted in flattening of caveolae ([Rothberg et al., 1992]).

5.3 Vesicles out of Cellular Plasma Membranes

One aim of the project was to study caveolae in a pure lipid membrane system, meaning independent of their cellular environment to discard any interaction with cellular compartments. First, a classical ("bottom-up") strategy was followed: the GST-cav1 and His-cav1 proteins were expressed in bacteria and purified with Gluathione agarose or a Ni-resin. In a second step, micelles and small unilamellar vesicles (SUV) of lipids, cholesterol and the purified cav1 were produced, from which giant unilamellar vesicles were formed via electroformation. Although the results of Lisanti and coworkers ([Li et al., 1996] could be reproduced in prooving the association of cav1 protein with the SUVs in using ultracentrifugation of a sucrose gradient (see Fig 5.3), there was no evidence that the cav1 protein was still present in the electroformed GUVs, and that they formed functional caveolae.

A posteriori, this unsuccesssful attempt is not surprising since cavin1 was shown to be a caveolin partner protein that is necessary to formation of budded caveolae ([Hill et al., 2008]). This led to a change in the strategy to obtain cytoskeleton-free model systems. In this new ("top-down") approach, the idea was to creata plasma membrane vesicles from the cell itself. Several protocols were developed to create vesicle-like structures from cell itself:

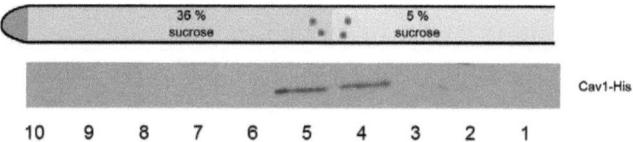

Figure 5.3: Immunoblot with a cav1-AB of cav1 in SUVs loaded on a sucrose step-gradient (fractions 1 : 40%, 2 − 4 : 32%, 5 − 10 : 5% sucrose). It is known that SUVs localize at the sucrose concentration step (fractions 4 and 5), and positive cav1 labeling of these fractions indicate association of cav1 to lipids in SUVs.

- The group from R. Scott produced giant plasma membrane vesicles (GPMV) using a buffer containing DTT or NEM ([Scott, 1976], [Holowka and Baird, 1983], [Baumgart et al., 2007])

- H. Vogel and coworkers used cytochalasin D to produce blebs ([Pick et al., 2005])

- The K. Simons group produced plasma membrane spheres (PMS) using simple PBS^{++} supplemented with glucose and Hepes ([Lingwood et al., 2008])

In Fig 5.3 a scheme of a bleb growing out of a cell is shown. All the above mentioned protocols were tested and are detailed below as they were used in our hands.

5.3.1 Giant Plasma Membrane Vesicles (GPMV)

Giant plasma membrane vesicles were produced from adherent cells following the protocol introduced by R. Scott ([Scott, 1976]). Briefly, confluent cells were washed twice in GPMV buffer ($150mM$ NaCl + $2mM$ $CaCl_2$ + $10mM$

induction of blebbing by incubation in bleb solution

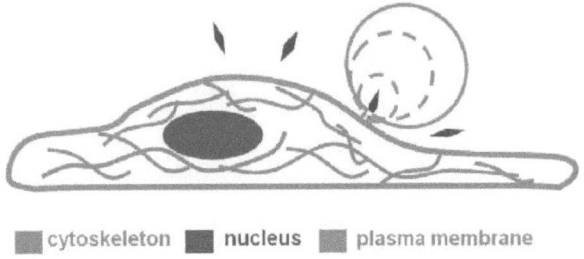

cytoskeleton nucleus plasma membrane

Hepes, all at $7,4pH$), then incubated in GPMV reagent (GPMV buffer + $2mM$ DTT + $25mM$ Formaldehyd) under continuous shaking ($\sim 80rpm$) at $37°C$ and 5% CO_2 for $40min$. The liquid was afterwards collected with a glass capillary ($1mm$ diameter), transferred into a $0,5ml$ Eppendorf and kept there for $10-40min$ to allow the vesicles to settle down. Vesicles could then be collected with a glass capillary from the bottom of the Eppendorf and transferred to the microscope stage for experiemtns.

5.3.2 CytochalasinD - Blebs

Incubation of adherent cells in PBS supplemented with $10\mu M$ cytochalasin D (CD) and $0,1\%$ BSA for $1h$ resulted in blebing cells. The blebs could be detached from the cells by shaking the dish, collecting the liquid with a glass capilary and transferring it into an $1,5ml$ Eppendorf. After $10-40min$ of settlement, vesicles could be collected for experiment from the bottom of the Eppendorf.

5.3.3 Plasma Membrane Spheres (PMS)

Plasma membrane spheres were obtained from adherent cells following a modified protocol, which was introduced by K. Simons ([Lingwood et al., 2008]). Briefly, confluent cells were washed twice in PBS and then incubated for $6 - 10h$ in PMS-buffer (PBS + $1,5mMCa^{2+}Cl_2$ + $1,5mMMg^{2+}Cl_2$ + $10mM$ Hepes) at $37°C$ and $5\%CO_2$. To improve the results, PMS-buffer was supplemented with $10\mu M$ of MG-132 (Calbiochem, Germany), which is a proteasome inhibitor ([Rock et al., 1994]) stopping the degradation of proteins such as caveolin. This mixture was called Cav-bleb-solution. PMS were seperated from the cells by shaking the dish, collection of the liquid with a glass capilary into an $1,5ml$ Eppendorf and transfer of vesicles into the microscope chamber after $40min$ of settlement. Another option was to run the PMS protocol on cells adherent on glass slides which afterwards were mounted directly on the microscope (**Epi-OT** or **Con-OT**)(Fig 5.3.3). Vesicles could then be detached from cells using a micropipette.

The presented results on cytoskeleton-free model systems were all obtained with plasma membrane spheres, because this technique provided the cleanest vesicles of a good size to work with.

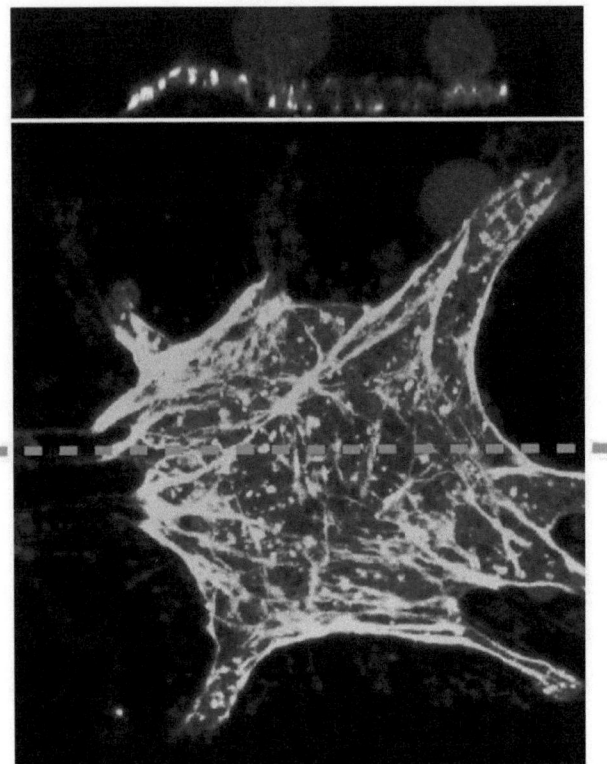

Figure 5.4: Confocal stack of a wt MLEC transfected with lifeact-mCherry after $6h$ of incubation in Cav-bleb-solution. Little amounts of cytosolic lifeact-mCherry fill the blebs attached to the cell, whereas it binds at high concentration to actin filaments in the cell body. Top: Profile of the cell along the dashed line (see below). Bottom: Projection of the stack showing the x-y-plane of the cell. Bar = $10\mu m$

Chapter 6

Experimental Set-Up

6.1 Tether Extraction

Two different set-ups were used for the extraction tethers: 1)an optical trap coupled into an inverted epi fluorescence microscope (in the following called **Epi-OT**) for experiments on cells and vesicles ([Cuvelier et al., 2005]), and 2) a confocal microscope equipped with an optical trap (in the following called **Con-OT**) for quantitative experiments on vesicles ([Sorre, 2010]). In both cases, the optical trap consisted of a single fixed laser beam focused by a 100X objective.

6.1.1 Epi-OT

The optical trap was created by steering a $1064nm$ laser beam (Coherent, Santa Clara, CA) into an inverted microscope (Axiovert 200, Zeiss, Germany), and was directed through a high numerical aperture objective (100X, 1.3 NA) via a hot mirror (Melles Griot, France) placed under the filter cube holder instead of a rotative analyzer (see Fig 6.1.1). This design allowed performing bright field or epi fluorescence imaging while trapping a bead. Analysis of thermal fluctuations of the bead revealed a trap stiffness of $K_{x,thermal}^{Epi-OT} = 131(\pm 6)\frac{pN}{\mu m \cdot W}$, and $K_{x,Stoke}^{Epi-OT} = 135(\pm 4)\frac{pN}{\mu m \cdot W}$ using Stokes law (6.1.9, page 94; [Bockelmann et al., 2002]). Bright field images were acquired with a CCD camera (XC-ST70CE, Sony, Tokyo, Japan) at 25 Hz, recorded on a computer with a video capture card (Piccolo Pro 2, Euresys, Angleur, Belgium) after contrast enhancement (Argus image processor, Hamamatsu Photonics, Hamamatsu, Japan). A mercury lamp served as the light source for epi fluorescence, and images were acquired with an EM-CCD camera (Cascade-CCD 97, Roper Scientific, GA, USA) controlled by Metamorph software (Molecular Devices, Ca, USA). The observation chamber containing cells was mounted on a temperature controlled stage of the microscope (Tempcontrol 37-2 digital, Carl Zeiss), and temperature was set to $37°C$ through the whole course of experiments (Fig 6.1.1).

6.1.2 Con-OT

A confocal microscope equipped with optical tweezers was used for experiments with fluorescently labelled vesicles, including micropipette aspiration and membrane tension measurement using tethers ([Sorre et al., 2009]). The possibility of visualization of the sample in bright field microscopy at video frame rate during confocal fluorescence acquisition was a special feature implemented by B. Sorre during his PhD (2007-2010).

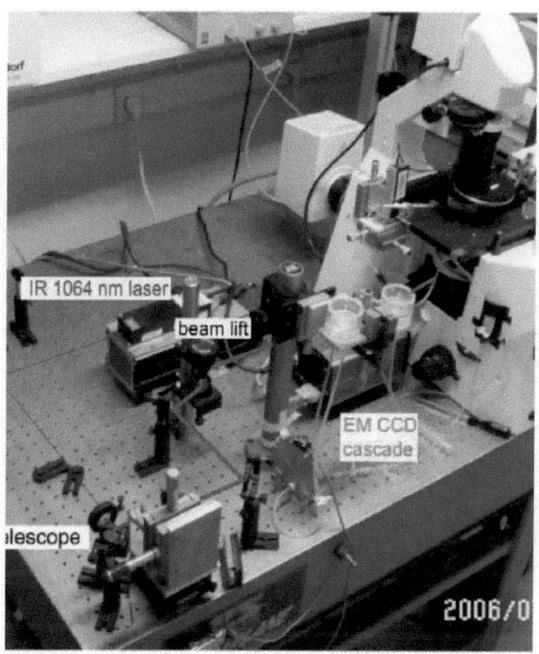

Figure 6.1: Set-up showing the alignment of the laser beam, which is broadened by the telescope and adjusted in height by two mirrors (beam lift) to enter the microscope. The position of the EM-CCD (Cascade) and the CCD camera are visible also.

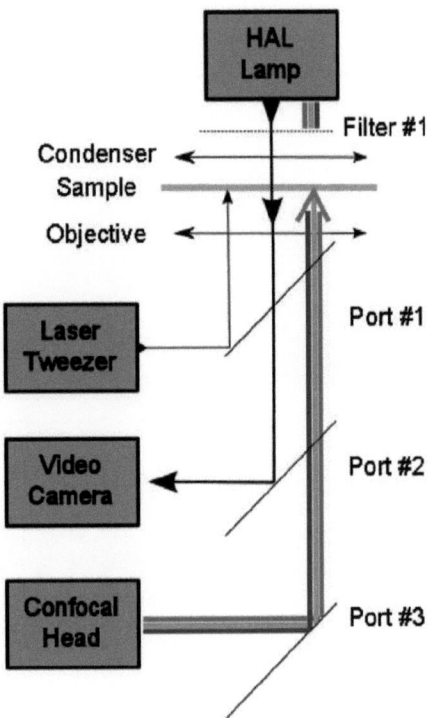

Figure 6.2: Schematic view of the experimental set-up. It shows the three independent channels used in the optical configuration. Confocal imaging uses visible light (port ♯3), optical trapping is achieved by a 1070nm continuous wave laser light (port ♯1), and bright field microscopy uses near infra red light (port ♯2). See also [Sorre, 2010]

The set-up was based on a commercial Nikon TE2000 inverted microscope (Nikon) modified with a stage riser (Nikon), in order to create an extra port (see port ♯1 in Fig 6.1.2). The confocal head was the eC1 confocal system (Nikon) equiped with two laser lines ($\lambda_1 = 488nm; \lambda_2 = 543nm$). Confocal microscopy, optical tweezers and bright field microscopy are not naturally compatible. Thus the light spectrum was split in three separate channels. Confocal fluorescence microscopy was performed in the visible ($400 < \lambda < 750nm$), optical trapping in the infra red channel ($\lambda > 900nm$) and finally bright field microscopy was operated in the near infra-red channel ($750 < \lambda < 900nm$). It was made using the fluorescence illumination arm as an imaging port. (for details see [Sorre, 2010]). Computation of the trap stiffness using Stokes law revealed $K_{x,Stoke}^{Con-OT} = 81(\pm 2)\frac{pN}{\mu m \cdot W}$ (a description of the calibration methods can be found below, 6.1.9, page 94).

6.1.3 Cell Stage and Pipette Holder

Since the optical tweezers were not movable, cells and vesicles had to be displaed in order to pull tethers with trapped beads. This controlled linear displacement was performed with a piezo-stage (PI, Karlsruhe, Germany). The **Epi-OT** setup (see [Cuvelier, 2005]) was first designed to work with vesicles, red blood, and other non-adherent objects, which were put into a small volume of adapted medium sandwiched between two glass slides connected to the cell stage. Then a micropipette connected to the piezo-stage and to an aspiration control system was inserted to manipulate single vesicles or cells (see Fig 6.1.3). This configuration was similar for the **Epi-OT** and

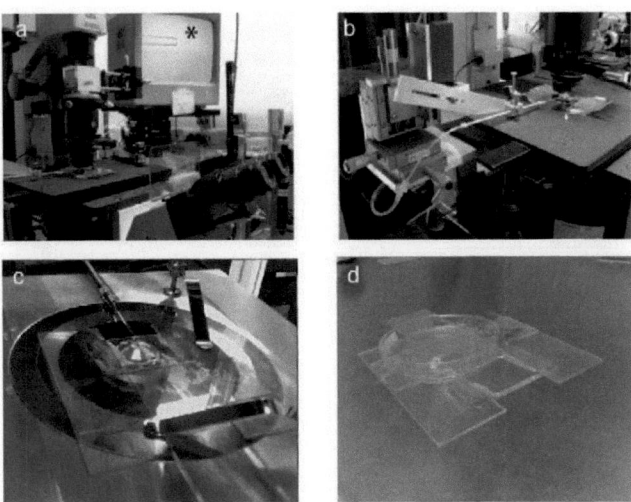

Figure 6.3: Pictures of the **Con-OT** set-up. (a) General view of the set-up. The microscope's field view is displayed on the TV-screen, where one ca distinguish the tip of the micropipette (⋆). (b) Zoom on the home-made three axis pipette manipulator. (c) Zoom on the pipette diving in the custom-made PMS-chamber. (d) Zoom on the custom-made PMS chamber.

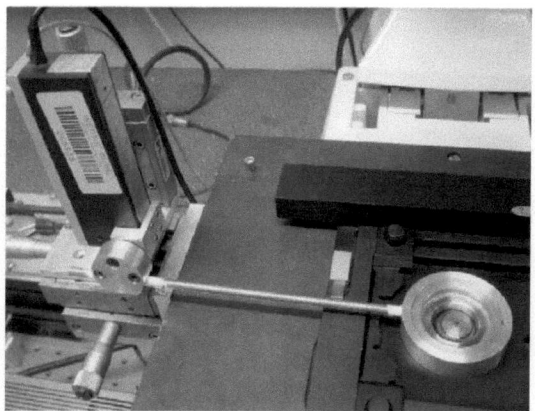

Figure 6.4: Picture of the home-made cell chamber connected to the piezo controlled micromanipulator via the "spoon"

the **Con-OT** set-up.

However, for adherent cells, the **Epi-OT** microscope could be modified. A custom-made metal cell chamber was placed into a specially designed holder (baptized "the spoon") connected to the piezo-stage. Aluminium allowed an efficient heat conductance from the temperature control stage to the cells. This allowed us to perform all cell experiments at 37°(Fig 6.1.2).

6.1.4 Hypo-osmotic Shock System

To apply a hypo-osmotic shock while measuring variations in the membrane tension, the switch from isotonic to hypotonic conditions had to be done smoothly. To do that, a needle was connected to the cell chamber, through which one fluid (e.g. H_2O) could be injected gently using a syringe. In addition, a second syringe connected to a needle was used to aspirate the

Figure 6.5: Picture of the custom-made chamber used for experiments with vesicles and micropipettes.

medium out of the chamber (see Fig 6.1.3). This home-made device allowed us to change media without changing the total volume present in the cell chamber. To avoid perturbing flows in the cell chamber center, small plastic brackets were placed in front of the needles.

6.1.5 Fabrication of Micropipettes

Micropipettes were made from borosilicate capillaries ($0.8mm$ inner/$1mm$ outer diameter). Capillaries were first pulled into fine cones using a laser pipette puller (P-2000, Sutter Instrument Co.). Then, they were cut open, and microforged at the desired inside diameter ($3 - 4\mu m$). The microforge consists of a glass bead heated by a titanium filament. The principle of the forge was based on the lower melting temperature of the glass bead compared

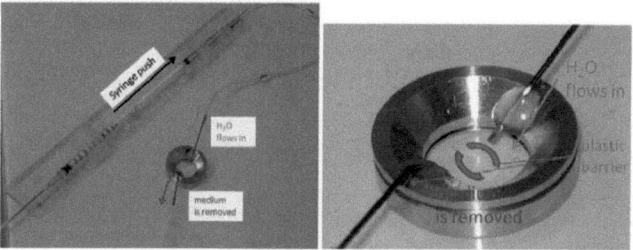

Figure 6.6: Left: Cell chamber with the hypo-osmotic shock system depicting the flow of liquids. Right: More detailed view of the cell chamber showing the connected needles and the position of the optional plastic brackets.

to that of the pipette. After the pipette was pushed into the melted bead, cooling resulted in a sharp break of the pipette. For the experiments, pipettes were filled with PBS, and attached to the chucks of the micropipette holder, which was connected to the micromanipulator. Further connection to the suction pressure device enabled aspiration control.

6.1.6 Aspiration Controll System

Two aspiration systems were used. The first was a simple syringe used to clean or unblock the micropipette. The second, which allowed a controlled aspiration, was based on the principle of communicating vessels. A water filled reservoir was connected to the back of the pipette by a silicone tube. This reservoir was mounted on a vertical rack, and its height was controlled by a mechanical translator (M105, Physik Instrumente)([Rawicz et al., 2000b])(Fig 6.1.6). The pressure difference at the entrance of the pipette between the observation chamber and the outside is determined by the height of the water reservoir Δh relative to the position of zero pressure:

$$P = \rho g \Delta h,$$

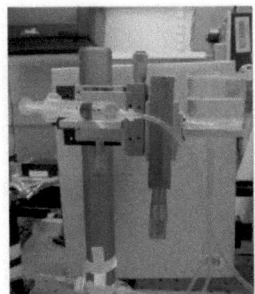

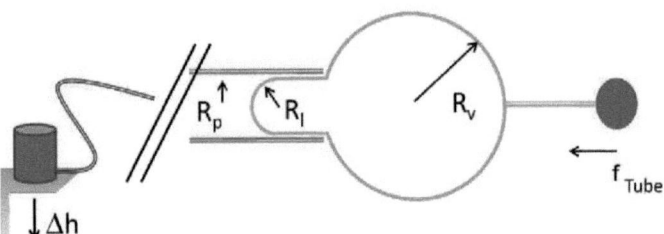

Figure 6.7: Top: The two aspiration systems; left: syringe, right: vessel installed on a rack. Bottom: Scheme of the vesicle aspiration with a micropipette connected to a water filled reservoir.

where ρ is the density of the medium in the reservoir ($\rho_{water} = 10^3 kg m^{-3}$), and $g = 9.81 \frac{m}{s^2}$ is the acceleration of gravity. The pressure used in the experiments ranges between $0,1$ and $100 Pa$ ($1 Pa = 1 \frac{kg}{m s^2}$). The surface tension of the membrane of an aspirated vesicle with radius R_v can be determined by applying the Laplace formula ([Kwok and Evans, 1981]):

$$\sigma = \frac{\Delta P R_p}{2(1 - \frac{R_l}{R_v})}, \quad (6.1)$$

where R_l is the radius of the cap of the aspirated membrane along the pipette axis (so called tongue) and R_p is the radius of the pipette.

6.1.7 Beads and Bead-coatings

For all experiments $3\mu m$ diameter polysterine beads (Polysciences Inc., PA, USA) covered with concanavalin A (Sigma, MO, USA) were used.

6.1.8 Online Tracking with MatLab

Bead tracking could be performed online, i.e. during the experiment, thanks to an algorithm based on Matlab (written by G. Toombes in the group). Briefly, a region of interest around the bead was chosen together with a threshold. The software was designed to compute the center of gravity of this region, which corresponded to the bead position (see Fig 6.1.8). The spatial resolution was $0,1 pixel$ or equivalent to a force resolution of $1 pN$ for the trap stiffness of $135 pN/\mu m$. It is noteworthy that the acquisition rate of 25 Hz was not limiting for the process that we have invesitgated.

6.1.9 Calibration

Trapping stiffness of the optical trap was calibrated by two methods. The fines one is based on the energy-dissipation theorem while the other is simply based on Stokes' viscous drag force ([Cuvelier et al., 2005]; [Bockelmann

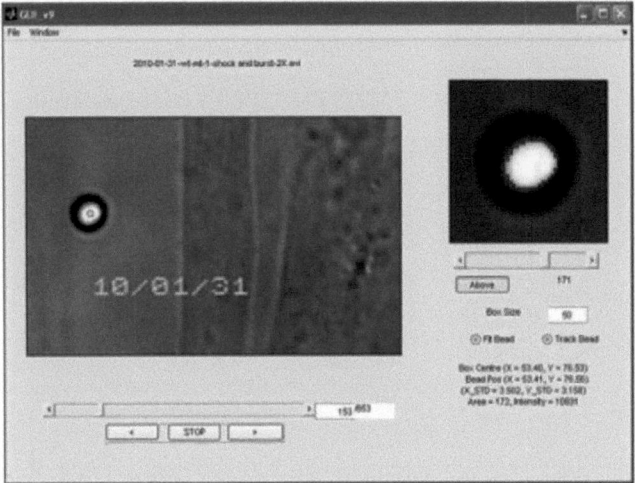

Figure 6.8: View of the interface of the tracking software. The small window on the upper right shows the region of interest, from which the center of gravity is computed.

et al., 2002]; [Svoboda and Block, 1994]).

Energy Dissipation Theorem Using the energy-dissipation theorem, the position fluctuations of a trapped bead can directly be related to the trap stiffness.

$$1/2 k_B T = 1/2 < \Delta r^2 > K, \qquad (6.2)$$

with $< \Delta r^2 >$ the mean square of the fluctuations and K the trap stiffness. Hence:

$$K = \frac{k_B T}{< \Delta r^2 >} \qquad (6.3)$$

This method is very precise for low laser powers, when thermal fluctuations are still big enough to be observed by video microscopy. At laser powers below $100 mW$, the stiffness K increases linearly with the laser power P_{laser} as shown in (6.1.9). At higher laser powers the trap becomes so stiff that the internal noise of the setup overwhelms the thermal noise of the bead. Assuming, that the linearity between the laser power and the trap stiffness is valid over the entire range of the laser power, we can use the slope obtained for the low power regime to calculate the trap stiffness for the laser power used in the experiment.

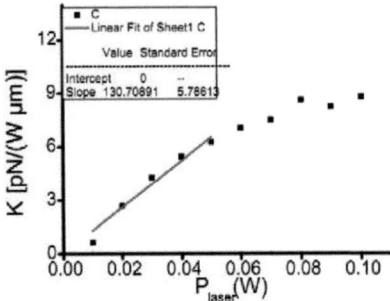

Figure 6.9: Calibration of the optical trap with thermal fluctuation. The slope in the linear regime $(0.01 - 0.06mW)$ is $K_P = 131(\pm 6)\frac{pN}{\mu m \cdot W}$. Above $0.08W$ the thermal fluctuations of the bead are overwhelmed by the noise of the data acquisition.

It is important to note that in general the shape of the laser beam in the focus is not perfectly circular. That leads to different trap stiffness in the x and y direction. We then need to define the bead displacement out of the center of the trap Δr by the cartesian coordinates Δx and Δy. Hence the total force exerted on the bead is composed of $f_{bead,x}$ and $f_{bead,y}$, i.e. the force in the x and y direction, respectively.

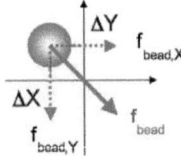

$$F_{bead} = \sqrt{F_{bead,x}^2 + F_{bead,y}^2} = \sqrt{K_x P_{laser} \Delta x^2 + K_y P_{laser} \Delta y^2}, \quad (6.4)$$

where $K_x P_{laser}$ and $K_y P_{laser}$ are the trap stiffness in x and y direction, respectively, for a given laser power P_{laser}.

Experimentally, the bead was only displaced along the x direction, and the force exerted on the bead is given by:

$$F_{bead} = K_x P_{laser} \Delta x \quad (6.5)$$

Stokes' Law Another method to calibrate the optical trap is based on the controlled displacement of the bead under external force. To do this, the fluid surrounding the bead is moved with a constant velocity v. In a fluid with the viscosity η, the force acting on the bead is given by the Stokes' formula:

$$F_{Stokes} = 6\pi\eta R v \quad (6.6)$$

The bead position is obtained by force balance $f_{trap} = f_{Stokes}$. Hence, for a given laser power and a fluid flow in the x direction, one finds

$$K = \frac{6\pi\eta R v}{\Delta x} \quad (6.7)$$

To apply a constant flow of fluid to the bead, the observation chamber was displaced by means of a piezo-electric element. To avoid shear effects due to surface interactions, the bead was positioned $3\mu m$ above the surface of the observation chamber. A triangular signal was applied to the piezo-electric element, with an amplitude corresponding to a displacement of $20-150\mu m$. The frequency of oscillation was fixed. The bead changed between the two equilibrium positions (separated by a distance of $2\Delta x$) following the direction of chamber displacement. Measuring the bead displacement at different laser powers P_{laser}, and distances of chamber displacement, allowed us to probe the trap stiffness in a wide range of laser powers as shown in (Fig. 6.1.9).

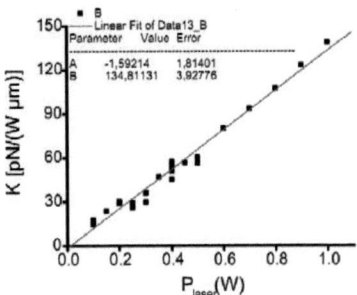

Figure 6.10: Calibration of the **Epi-OT** optical trap with chamber displacement. A linear fit of the data gives the slope $K_P = 135(\pm 4)\frac{pN}{\mu m \cdot W}$.

6.2 TIRF-microscopy

Total internal relection microscopy (TIRF) was performed with cells in order to visualize and to identify single caveolae at the plasma membrane. After a short introduction to the principles of this technique a description of our set-up will follow. For more detailed information about TIRF microscopy, see ([Axelrod, 2008]).

6.2.1 Principle of TIRF

TIRF microscopy is a method of choice to visualize fluorescently labeled objects in the vicinity of a surface while increasing the signal to noise ratio if the bulk phase also contains fluorescent species. TIRF microscopy was performed using a Zeiss Axiove rt 200 inverted microscope. The different wavelengths (488 nm, 568 nm, 614 nm) of an argon-krypton mixed gas laser (Melles Griot,Carlsbad, CA) were selected by an acoustic-optical tunable filter unit, routed to the microscope by a single-mode polarization-maintaining fiber optic and the TIR mode generated by illumination through the periphery of a high NA objectve ($100X$, $1, 45NA$). Images were acquired by an EM CCD camera ($8\mu m$ pixel dimension, $1000x1000$ pixels, C 9100-02, Hamamatsu Photonics, Japan). For dual wavelength imaging, the emission light was passed through a dual-view optical unit (Photometrics, Tucson, AZ) that splits the separate wavelengths to different parts of the CCD chip.

Part III

Results

At the onset of this work, it was investigated whether and how caveolae might play a role in cell tension regulation. The first part will be about the effct of caveolae on the membrane tension of cells at rest. Then the caveola mediated response to acute mechanical stress will be studied. Finally will be mentioned our latest results on minimal membrane systems (i.e. plasma membrane spheres) containing caveolae and on muscle cells (myotubes), which were aimed at establishing a possible link between caveolin-mutations and muscular dystrophies. But, let us start with explaining how a typical tether extraction experiment was performed.

Chapter 7

Tether Extraction From Adherent Cells

7.1 Typical Tether Force Traces

A typical tether extraction experiment will be described in this section. Unless otherwise stated, the results presented in the rest of the manuscript were obtained following the same experimental procedure.

Fig 7.1 shows a representative force-time trace during a typical tether extraction from an adherent cell (here a wt MLEC). This process is composed of 5 phases:

I After a cell was chosen for an experiment, the trapped bead was positioned in front of the cell, and the image acquisition was started in order to capture the position of the fluctuating bead in the absence of a membrane tether. This position sets the force reference ($f = 0$) via the trap stiffness ($0 - 10$ seconds in Fig 7.1)

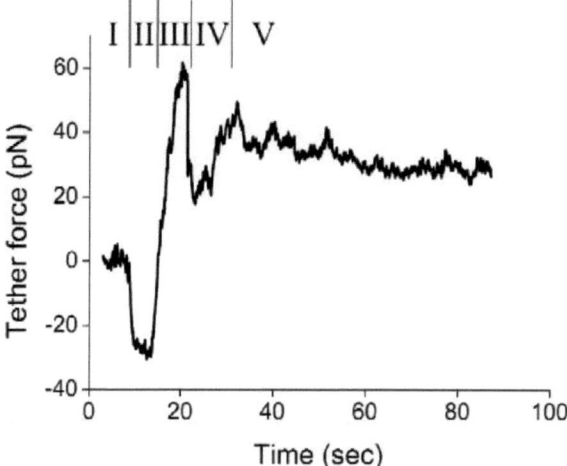

Figure 7.1: Top: Video screenshots of the experiment showing the bead (center) and the cell (right) at different steps of the tether extraction. Bottom: Corresponding force-time trace; the roman ?gures indicate the different steps of tether extraction: (I) Reference force corresponding to the position of the free, untethered bead. (II) Negative force corresponding to impingement between bead and cell to establish an adhesive contact. (III) Tether nucleation. (IV) Tether elongation at pulling velocity $v \sim 0.5 \mu m/s$ (final length $\sim 10 \mu m$). (V) Tether force relaxation down to the static force.

II The bead was brought into contact with the cell by moving the cell towards the bead. This pushed the bead out of its reference position (corresponding to the negative force in Fig 7.1). Moreover, the bead was lowered in z position in order to have a broader contact site with the cell membrane. A contact time of $2-20$ seconds was typically necessary to establish an adhesive link between cell membrane and bead, depending on the cell type, the cell treatment, and the freshness of the bead coating.

III After a few seconds of contact, the cell was moved away from the bead slowly (a few μm). An adhesive contact between the bead and the cell membrane resulted in visible displacement of the bead in the direction of cell displacement (positive force increase in Fig 7.1) until a tether was nucleated (high peak). This force overshoot is the signature of tether formation. As mentioned in 1.2.2 (page 24) the magnitude of the force peak depends on the size of the adhesive contact ([Koster et al., 2005]).

IV To ensure that only one single tether remained, the tether was further elongated at constant speed ($v \sim 0.5 \mu m/s$), which resulted in a slow increase of the tether force. In the case of multiple tethers, the force increase would be much steeper, and a broader "thread" (or bundle of tether) would be easily. Eventually the rupture of single tethers with the corresponding steps in the tether force would be observed.

V At $t = 30$ seconds, elongation is stopped. The tether force is observed to relax to a plateau value. The static force was computed as the mean value of the tether force measured during at least 10 seconds after relaxation.

Chapter 8

Preliminary Remarks and Comments on the Relation Between Tether Force and Membrane Tension on Cells

The force overshoot associated to tether nucleation will be overlooked in this work, although it is a particular feature in the force trace. It is mostly dependent on the protocol details of force application (size of adhesion contact, controled by contact time, impingement force, bead coverage...). Our initial goal was to directly test the Sens and Turner model ([Sens and Turner, 2005]) and use the tether pulling technique as a way of both, extracting membrane area in a controle d manner (and thus applying a local perturbation to the cell membrane), and measuring the resulting cell tension. However, as seen during phase IV of tether extrusion and already reported by others

([Sheetz, 2001], [Heinrich et al., 2005]), elongation of tethers extracted from living cells significantly deviates from the behavior discussed for lipid bilayers. Instead of exhibiting a length-indep endent plateau value, the tether force increases up on elongation even at small pulling velocities as small as fractions of $1\mu m/s$. This process very likely originates from viscous friction between cytoskeleton and membrane, and from the flow of membrane around transmembrane proteins. A comprehensive physical mechanism of this dynamic contribution to the tether force is unde rcurrent investigation since the force velo city relationship is still debated. In order to avoid the complexity inherent to dynamics of tether extrusion, it was decided to focus on the static tether force, as measured after relaxation at constant length. As a consequence, our tether pulling assay will only serve as a sensitive technique to probe the tension of the cellular interface. In the following chapter, it will be first investigated whether the presence of caveolae at the cell plasma membrane directly contributes to the resting cell tension. Then, by externally applying mechanical stresses (such as osmotic shocks), variations of tether forces will be measured to assess the cell membrane response under stress. Finally, it is important to recall that measurements of tether forces yield the cell tension, which is not the strictly the membrane tension of the cell. As mentioned earlier in 1.2.2 (24) the force required to hold a tether extracted from a lipid bilayer is a direct measure of the membrane tension σ (providing that the bending rigidity is known). Contrastingly, in the case of living cells, the same formalism can be conserved if σ is replaced with the effective cell tension, which is the sum of the plasma membrane tension σ and the cytoskeleton-membrane adhesion energy density W_0. Moreover, Sheetz and coworkers have shown that $\sigma \ll W_0$ in eukaryotic cells with intact actin cortex ([Raucher et al., 2000]). In consequence, under identical experimental and physiological conditions, differences in tether forces measured between a wt and a mutant cell line will be mainly informative about how the mutation alters W_0. The best way to gain insight into the tension of the plasma membrane itself is to investigate how the tether force is affected by changes in experimental conditions that do not signifficantly impair the cytoskeleton-membrane adhesion. As abovementioned, the caveola-mediated cell response under hypo-osmotic shock will be studied, which may seem to be a harsh treatment that also influences W_0. However, at this stage , it is important to keep in mind that our focus is on the short-term cell tension regulation. It is thus assumed that hypo-osmotic shocks do not signifficantly affect W_0 within the first 5 minutes following the shock. This assumption has already been made in previous works by Sheetz and coworkers ([Dai et al., 1998]) and is supported by the fact that the morphology of the cytoskeleton is not observed to be altered within this short 5 minutes duration ([D'Alessandro et al., 2002]).

Chapter 9

Caveolae and the Resting Cell Tension

In this chapter, two cell lines were selected, namely MLEC and MEF. In both cases, the effective cell tension of cells rich in caveolae (wt) and cells devoid of caveolae (cav $1^{-/-}$) will be compared in resting physiological conditions. It will be also investigated how chemical and biological treatments impact on the measured tether force.

9.1 The Effective Tension of MLEC is Affected by the Presence of Caveolae

The effective cell tension of mouse lung endothelial cells in resting (physiological) conditions was measured in extracting membrane tethers with optically trapped beads as described in the previous section. The histogram in Fig 9.1 shows the distribution of the tether force for both wt and $cav^{-/-}$ MLECs.

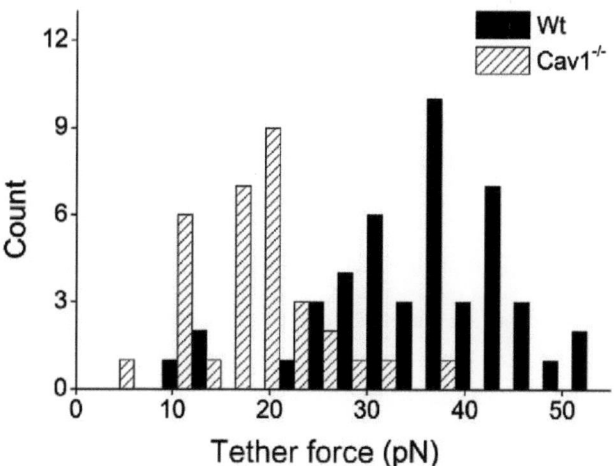

Figure 9.1: Histogram of static tether force measuremnts on wt and Cav1$^{-/-}$ These data correspond to different tethers pulled from different cells.

For both cell lines, the mean of the distribution was computed. As it can be seen in Fig 9.1 there is a significant difference in the teter force between the two cell types. In wt MLEC, the tether force was $f^{wt} = 35 \pm 1.4pN$ ($N = 46$), whereas it was $f^{cav1-/-} = 18 \pm 1.2pN$ ($N = 32$) for cav1$^{-/-}$ MLEC.

Transfection of cav1$^{-/-}$ cells with EGFP-cav1 restored the tether force to $f^{EGFP-cav1} = 38 \pm 1.8$ ($N = 34$) in cells showing a caveolar pattern of EGFP as observed with epi fluorescence microscopy (Fig 9.1). This indicates that the presence of caveolae indeed accounts for the measured difference in tether force between wt and cav1$^{-/-}$ cells. It is for note that this twofold difference

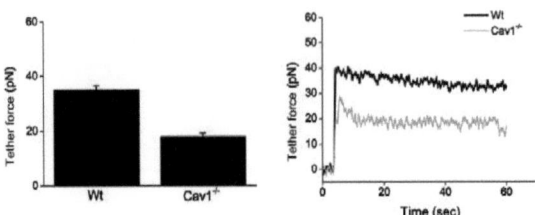

Figure 9.2: Left: Plot of the tether force in wt ($N = 46$) and cav1$^{-/-}$ ($N = 32$) MLEC. Data represent mean ± standard errors. Right: Example of single experiments showing the static tether force of a wt (black) and cav1$^{-/-}$ (grey) MLEC.

in force means that wt and cav1$^{-/-}$ cells are characterized by effective resting tension that differs by 4-fold (according to Eq. 1.18).

9.2 The Effective Tension in MEFs Does not Depend on Presence of Caveolae

To check whether the effect of caveolae on the cell tension observed in the MLEC was a general trend, tether force measurements were also performed on mouse embryonic fibroblasts (MEF). As for MLEC, two MEF cell lines were used: the caveolae containing wt form and the cav1$^{-/-}$ variant. By contrast with MLEC, the measured tether forces in wt and cav1$^{-/-}$ MEF did not exhibit any significant difference ($f^{wt\,MEF} = 11 \pm 1.7\,pN$, $N = 4$, and $f^{cav1^{-/-}\,MEF} = 11 \pm 1.4\,pN$, $N = 4$; Fig 9.2), meaning that the effective cell tension is not affected by the presence of caveolae.

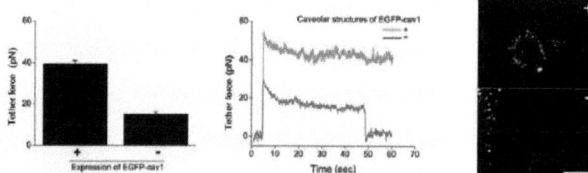

Figure 9.3: Left: Plot of the tether force in cav1$-/-$ MLEC transfected with cav1-EGFP showing protein expression (+) or not (-). Data represent mean ± standard errors. Middle: two single experiments showing the static tether force of cav1-EGFP expressing cav1$^{-/-}$ MLEC with caveolar like structures (+) or not (-). Right: Fluorescence images of cav1-EGFP corresponding to the cells in the middle panel.

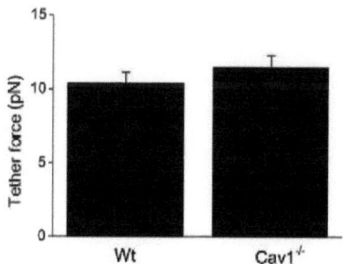

Figure 9.4: Plot of the tether force in wt and cav1$^{-/-}$ MEF. Data represent mean ± standard errors.

9.3 Challenging the Effective Cell Tension by Chemical and Biological Treatments

To gain better insight into the possible contribution of caveolae to the effective resting tension of cells, another series of experiments was performed, in which cellular components were disturbed by either chemical or biological treatments. The cytoskeleton was altered by several drugs like cytochalasin D (CD), latrunculin A (LatA), and jasplakinolide (Jas). The cell was also depleted of energy (i.e. ATP), the plasma membrane was depleted of cholesterol by m-β-cyclo dextrin (mβCD) which is known to result in the flattening of caveolae ([Rothberg et al., 1992]), and the activity of *src*-kinase was hindered by the phosphatase inhibitor PP2. Additionally, cells were transfected with fluorescently labeled caveolins displaying point mutations (cav1-Y14F-YFP, cav3-P104LL-YFP, and cav3-GFP).

9.3.1 Alteration of the Cytoskeleton Decrease the Effective Cell Tension

The tether force, and thus the effective cell tension of both wt and cav1$^{-/-}$ MLEC decreased to similar values after treatment with the abovementioned drugs disturbing the actin cytoskeleton (Fig 9.3.1), more precisely, when cells were incubated for 10-20 minutes in

- $0.5\mu M$ **CD**, the mean tether forces were found to be $f^{wt,CD} = 10.8 \pm 0.9\,pN$ ($N = 22$) in wt, and $f^{cav1^{-/-},CD} = 8.5 \pm 0.8\,pN$ ($N = 27$) in cav1$^{-/-}$ cells

- $1\mu M$ **LatA**, $f^{wt,LatA} = 8 \pm 1.7\,pN$ ($N=6$) in wt, and $f^{cav1^{-/-},LatA} = 6.0 \pm 1.0\,pN$ ($N=8$) in cav1$^{-/-}$ cells

- $1\mu M$ Jas, $f^{wt,Jas} = 16.0 \pm 0.8\,pN$ ($N=26$) in wt, and $f^{cav1^{-/-},Jas} = 11.3 \pm 0.9\,pN$ ($N=18$) in cav1$^{-/-}$ cells.

The three different drugs perturbed the actin cytoskeleton in different manners, but in all case, the treadmilling of actin filaments was impaired. The direct consequence was that the effective cell tension significantly decreased as expected ($W_0 \sim 0$). The measured value was similar for wt and cav1$^{-/-}$ cells.

9.3.2 ATP Depletion Decreases the Membrane Tension

After 50 minutes of ATP depletion, the static tether force was measured to be $f^{wt,noATP} = 3.8 \pm 0.5\,pN$ ($N=12$) in wt MLEC, and $f^{cav1^{-/-},noATP} = 5.0 \pm 0.5\,pN$ ($N=8$) in cav1$^{-/-}$ cells (Fig 9.3.2). Upon ATP depletion, the effective cell tension thus decreased, irrespective to the presence or absence of caveolae.

9.3.3 Interaction of Cav1 with Src-kinase

One of the interaction partner with cav1 involved in the formation and endocytosis of caveolae are *src*-kinase proteins which phosphorylate tyrosine 14 of the cav1 protein. Inhibition of *src*-kinase was reported to reduce the new formation of caveolae after the addition of epithelial growth factor ([Orlichenko et al., 2006]). Two strategies were chosen to address the influence of cav1-tyrosine 14 phosphorylation on the cell tension.

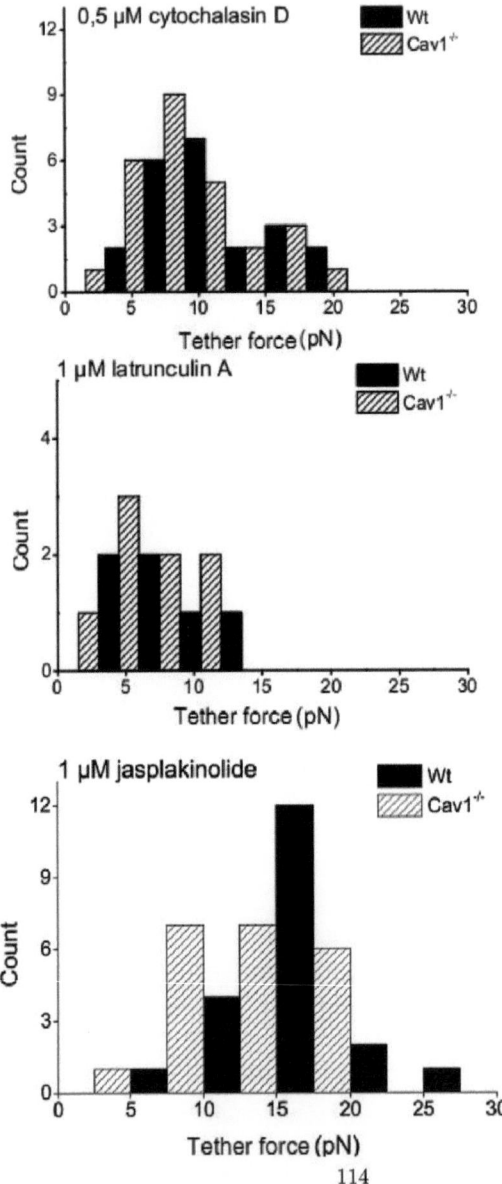

Figure 9.5: Histograms of static tether force measurements on wt (black) and cav1$^{-/-}$(striped) MLEC after 10-20 minutes treatment with $0.5\mu M$ CD (top), $1\mu M$ LatA (middle) or $1\mu M$ Jas (bottom).

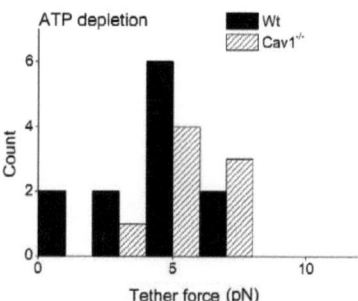

Figure 9.6: Histogram of static tether force measurements on wt (black) and cav1$^{-/-}$ (striped) MLEC after 50 minutes of ATP depletion.

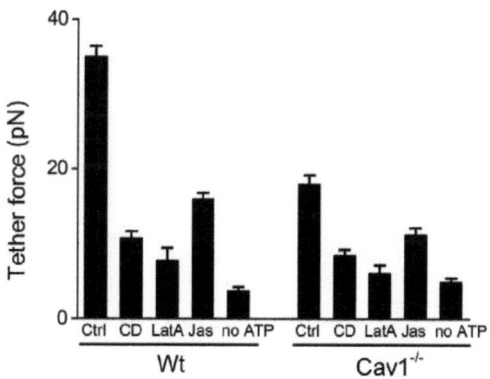

Figure 9.7: Plot showing the tether forces of wt (left) and cav1$^{-/-}$ (right) MLEC in control conditions (Ctrl), with $0.5\mu M$ CD, $1\mu M$ LatA, $1\mu M$ Jas or after ATP depletion (no ATP). Data represent mean ± standard errors.

- wt MLEC were pretreated with $20\mu M$ PP2 to inhibit the activity of *src*-kinase

- cav1$^{-/-}$ MLEC were transfected with the mutant cav1-Y14F-GFP which cannot to be phosphorylated.

PP2 treatment of wt MLEC did not alter significantly the cell tension, since the static tether force remained at $f^{wt;PP2} = 35 \pm 3.7\,pN$, $N = 6$). Similarly, cav1-Y14F-GFP was able to restore the membrane tension of cav1$^{-/-}$ MLEC, when it was expressed sufficiently high ($f^{cav1^{-/-},Y14F,+} = 38 \pm 2.7\,pN$, $N = 6$), whereas the membrane tension of cells with low expression remained at cav1$^{-/-}$ levels ($f^{cav1^{-/-},Y14F,-} = 18 \pm 2.0\,pN$, $N = 8$)(Fig 9.3.3). These findings suggest that the resting tension of MLECs is solely governed by caveolae present at the membrane. The sole expression of cav1 is not sufficient to re-establish the cell tension or the formation of new caveolae.

9.3.4 Cav3 Re-establishes the Cell Tension of Cav1$^{-/-}$ MLEC

Cav1$^{-/-}$ MLEC were transfected with cav3-EGFP to check the specificity of the observed effect of caveolae formed by cav1 on the cell tension. As observed in epi-fluorescence, transfected cells showed the same membrane pattern of cav3-EGFP as the cav1-EGFP. Moreover, the effective tension of these cells was restored to the level of wt MLEC, as revealed by tether extraction experiments ($f^{cav3-EGFP} = 34 \pm 1.9\,pN$, $N = 13$). Thus, cav3, which is the natural caveolin isoform for muscle cells, is also able to form caveolae in MLEC, and these caveolae have similar effects on the resting

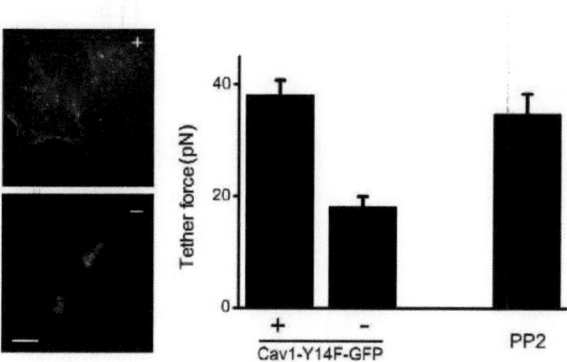

Figure 9.8: Left: Epi fluorescence imaging of cav1^{-}/$-$ MLEC expressing high (+) or low (-) levels of cav1-Y14F-GFP. Bar = $10\mu m$. Right: Plot showing the mean tether forces of cav1$^{-/-}$ MLEC expressing high (+) or low (-) levels of cav1-Y14F-GFP, and of wt MLEC pretreated with PPS (PP2). Data represent mean ± standard errors.

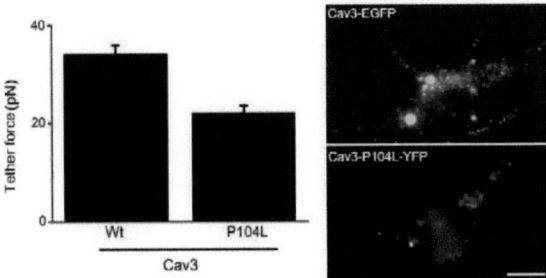

Figure 9.9: Cav1$^{-/-}$ MLEC transfected with cav3-EGFP and cav3-P104L-YFP. Left: Plot showing the mean tether forces of cells transfected with cav3-EGFP (wt) and cav3-P104L-YFP (P104L). Data represent mean ± standard errors. Right: Fluorescence images showing the localization of cav3-EGFP and cva3-P104L-YFP, respectively.

cell tension. Cav1$^{-/-}$ MLEC were also transfected with the point mutation cav3-P104L (analogue to the cav1-P132L mutation), which is a dominant negative mutation unable to exit the Golgi apparatus. Within transfected cells, not any could be found with a caveolar plasma membrane pattern of cav3-P104L-YFP. As expected, the effective tension of these cells remained at the level of cav1$^{-/-}$ MLEC ($f^{cav3-P104L} = 22\pm1.5\,pN$, $N = 12$)(Fig 9.3.4).

9.4 Summary

The main findings can be summarized as follows:

- The presence of caveolae in untreated MLECs increases the effective cell membrane tension as compared to MLECs devoid of caveolae.

- This difference vanishes in cells that exhibit an altered (decreased) cytoskeleton-membrane interaction (i.e. decrease in adhesion energy density).

- The influence of caveolae on the effective cell membrane tension is cell line dependent, since wt and cav1$^{-/-}$ MEFs are characterized by the same cell membrane tension.

These results suggest that caveolae do not per se affect the resting tension of the cell plasma membrane. Instead, their presence may have a detectable influence on the cytoskeleton-membrane adhesion, as revealed by our data on MLECs. This finding is consistent with the fact that caveolae directly or indirectly interact with the actin machinery, as mentioned in 3.1 (page 42). Yet, this trend cannot be generalized to all cell lines (see e.g. MEF). The presence of caveolae or their suppression may have indeterminate effects on W_0. The interaction of caveolae with the actin cortex may thus be enhanced or compensated by other alterations of the cytoskeleton-membrane interaction due to genetic mutations. This intricate effect is difficult to predict.

Chapter 10

Caveola-mediated Membrane Tension Buffering Upon Acute Mechanical Stress: Experiments on Cells

In this chapter, it is investigated how caveolae may contribute to the cell response upon mechanical stress. The results presented hereafter are the core of a published article ([?]).

10.1 Application of Acute Mechanical Stress and Cell REsponse Observed by TIRF and EM

Several ways exist to expose cells to an acute mechanical stress like

- shear stress induced by liquid flow

- uniaxial stretch induced by elongation of elastic cell substrates, on which cells are plated

- swelling upon application of hypo-osmotic shock.

The application of hypo-osmotic shock was chosen mainly, because it allowed us to easily combine the application of stress with tether extraction. To demonstrate that the results described below were not specific to osmotic shocks, the fate of caveolae up on b oth hyp o-osmotic sho ck and uniaxial stretch was studied by TIRF microscopy in cav1-EGFP expressing cells (experiments performed by B. Sinha). The results obtained with both kinds of mechanical stress are comparable, which supports the physiological relevance of osmotic shock application at short time scales ($\sim$ 5 minutes).

10.1.1 Mechanical Stress Leads to the Partial Disappearance of Caveolae from the Plasma Membrane

First the effect of the hypo-osmotic conditions on the cell volume was tested with HeLa-PFPIG cells (expressing cav1-EGFP). Switching from an osmolarity of $300mOsm$ to $30mOsm$ resulted in an immediate increase of cellular volume visualized by 3D confocal microscopy. Quantitative image analysis further showed that the cell volume increased by about 35%, peaking within the first 5 minutes. The cell volume was then observed to decrease slowly while hypo-osmotic conditions were maintained. Moreover, on reversing the external osmolarity back to $300mOsm$ after 30 minutes of hypotonic shock, the volume decreased to a value below the initial cell volume. These two latter observations support the existence of a compensatory cellular mechanism known as regulatory volume decrease, which originates from the enhanced activity of ions channels to restore the osmotic balance, as previously reported ([D'Alessandro et al., 2002]). The data however suggest that this process is not dominant during the first 5 minutes following hypo-osmotic shock. (Fig 10.1.1 A, B). TIRF microscopy was used to visualize plasma membrane caveolae. A dedicated image analysis platform enabled to follow the dynamics of caveolae upon hypo-osmotic shock (LabView program written by B. Sinha).

Upon switching the osmolarity from $300mOsm$ to $30mOsm$, the first striking observation was that the number of caveolae present at the cell surface of individual cells significantly decreased by about 30% at the surface of individual cells within minutes of hypo-osmotic shock (Fig 10.1.1 C,D). This effect was not due to a detachment of the cell during the hypo-osmotic shock, as it was checked by Reflection Interference Contrast Microscopy (RICM) (Fig 10.1.1).

Then, a stretching device was designed by B. Sinha and build by the Curie-workshop. This device was based on the use of a thin transparent silicone substrate, and was compatible with TIRF microscopy (Fig 10.1.1). This set-up enabled high quality imaging of caveolae dynamics by TIRF in a given

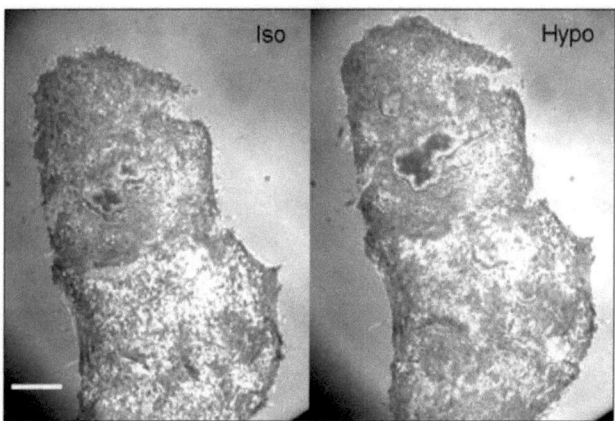

Figure 10.1: RICM image of cells before (Iso) and after hypo-osmotic shock of 5 minutes (Hypo). Upon hypo-osmotic shock, the proximity of the plasma membrane to the glass surface is increased (expansion of the central dark patch) ruling out the possibility of loss of contact between the plasma membrane and glass surface. Bar = $5\mu m$.

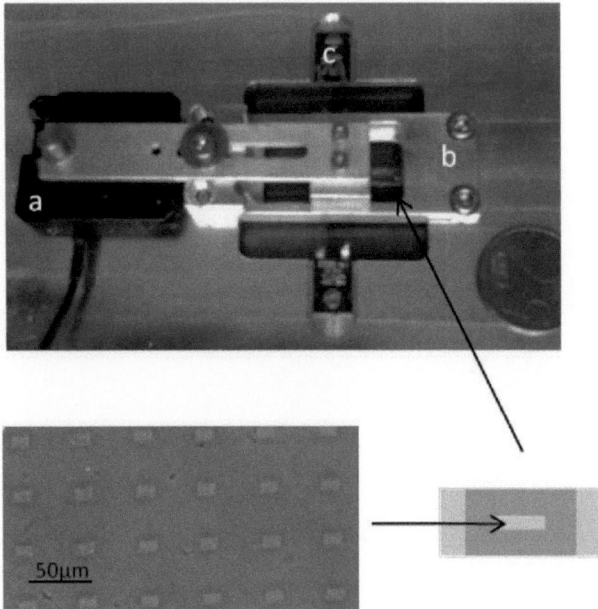

Figure 10.2: Top: Snapshot of the stretching device with the following components marked out: a) linear actuator for pulling b) detachable mount for PDMS sheet with cells c) resistors for temperature control. Bottom: Phase contrast image of micro patterned adhesive patches on PDMS for regulating cell shape and orientation. PDMS sheets ($100\mu m$ thick) thus micro patterned (light blue) are joined to a thicker ($1mm$ thick, shown as darker blue) sheet with a central window, so as to make a chamber for plating cells. The chamber is then assembled on the stretching device. provided by B. Sinha

cell before and after stretch. Additionally, to allow quantitative analysis, the use of stretchable substrates was combined with micro patterning technology ([Chen and Norkin, 1999]) to control the adhesion area and geometry of the cell as well as its orientation with respect to the stretching axis. As shown on Fig 10.1.1 F and G, the number of caveolae present at the basal footprint of cav1-EGFP HeLa cells was decreased up on stretching, and the loss of caveolae correlated with the extent of cell stretching (Fig 10.1.1 H). Remarkably, caveolae disappearance follows a similar pattern upon hypo-osmotic shock or membrane stretching (Fig 10.1.1 E and H)

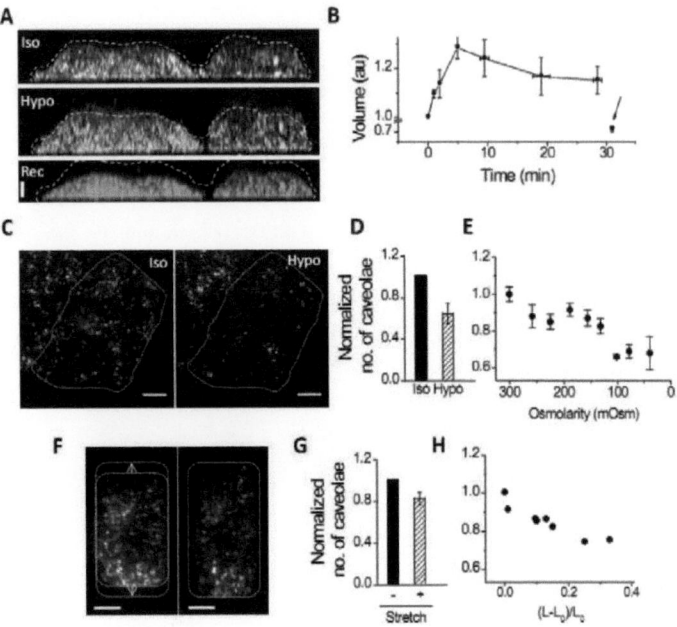

Figure 10.1.1 (A) Hypo-osmotic shock causes cell swelling. YZ maximum-intensity projection of confocal stacks of HeLa cells stably expressing cav1-EGFP. Top panel is the projection of 4 cells under iso-osmotic conditions (Iso), middle panel shows the same cells under hypo-osmotic conditions (Hypo) 5 minutes after switching osmolarity from 300 to $30mOsm$, and the bottom panel shows the cells 3 minutes after returning to $300mOsm$ (Rec) subsequently to hypo-osmotic shock for ~ 30 minutes. Bar = $50\mu m$. Dashed lines mark out the cell boundary for the iso-osmotic condition. (B) Volume of HeLa cells stably expressing cav1-EGFP tracked from iso-osmotic conditions to different time points after hypo-osmotic

shock ($30mOsm$, $t = 0min$) and upon returning to iso-osmolarity ($300mOsm$, $t \sim 29$ minutes). Arrow indicates the volume measured upon recovering iso-osmolarity. Data derived from multiple measurements ($N = 5$) in 3 independent experiments. Error bars represent standard deviations. (C) TIRF images of HeLa cav1-EGFP cells under iso-osmotic conditions (Iso) and after hypo-osmotic shock (4 min) (Hypo). Dotted line marks out the cell footprint. $Bar = 50\mu m$. (D) Change in the number of caveolae for single HeLa cav1-EGFP cells after hypo-osmotic shock (Hypo) normalized to the number counted before hypo-osmotic shock (Iso) ($N = 18$). Error bars represent standard deviations ($p = 4E^{-11}$). (E) Evolution of the loss of caveolae per cell with decreasing osmolarity. The same HeLa cav1-EGFP cells were exposed to decreasing osmolarities during ~ 1 minute for each osmolarity. Error bars represent standard deviations ($N = 3$). Data are representative of 3 independent experiments.(F) TIRF images of a single cav1-EGFP HeLa cell on the stretching device at 0% (left), and 20% stretch (right). Dotted lines mark out cell boundaries before and after stretch. Note that transverse contraction is negligible. $Bar = 50/mum$. (G) Change in number of caveolae for single HeLa cav1-EGFP cells after stretching ($15 \pm 1\%$) normalized to the number counted before stretching ($N = 7$). Data derived from multiple measurements ($N = 7$) in 7 independent experiments ($p = 0.00033$). (H) Evolution of the number of caveolae for single HeLa cav1-EGFP cells stretched to different lengths. Stretching was calculated from the change in the lateral dimension of the cell footprint in the direction of the stretch as $(L - L_0)/L_0$ where L_0 was the initial length and L was the ?nal length of the cell footprint. Each point represents the loss of caveolae for a particular cell. The number of caveolae is found to be negatively correlated to the amount of stretch ($N = 7$, $r^2 = 0.85$) as measured in 7 independent experiments.

10.1.2 Partial Disapperance of Caveolae Observed by EM

Electron microscopy (EM) studies were performed on wt MLEC at iso-osmotic conditions, and upon hypo-osmotic shock, to study on the ultra structural level the fate of caveolae upon acute mechanical stress. Two EM techniques were applied, 1) EM of ultrathin cryosections combined with gold particle immuno-labeling showing lateral cuts of cell membrane parts (in collaboration with G. Raposo), and 2) deep-etched EM showing the structure of the basal cell membrane adherent to the cover slide viewed from inside the cell ([Morone, 2010], in collaboration with N. Morone). While the fraction of lost caveolae quantified by EM on ultrathin cryosections was in agreement with values obtained by TIRF, deep-etched EM allowed us to observe "ghosts" of flat caveolae (Fig 10.1.2).

10.2 Membrane Tension Measurements During Hypo-osmotic Shock

In parallel with the visualization of caveolae upon cell exposure to hypo-osmotic shock or stretch (EM and TIRF), we used tether extraction experiments during hypo-osmotic shock to study the evolution of the cell tension, which is likely to reflect the variations of the membrane tension (see Eq. (1.18)). As discussed in chapter 8, the effective cell tension of cells measured in iso-osmotic conditions significantly varied with the cell type (MEF, MLEC) and chemical treatments (ATP depletion, actin drugs, cholesterol

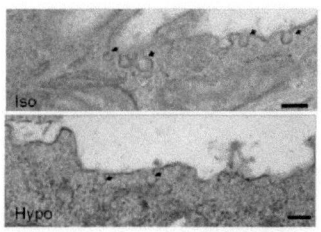 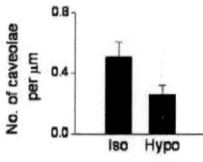

Figure 10.3: Left: Ultrathin cryosections of wt MLEC before (Iso) and 5 minutes after (Hypo) switch to hypo-osmotic medium ($30mOsm$) examined by electron microscopy. Right: Quantification of caveolae detected per μm of plasma membrane on ultrathin cryosections of MLEC in iso- or hypo-osmotic conditions reveal a signi?cant decrease in the number of caveolae per mum after hypo-osmotic shock. Total membrane used for quantification was 76 and 67μm for iso- and hypo-osmotic conditions, respectively. Data represent mean $\pm$ standard errors ($p = 0.047$).

extraction), and differences may also arise from the presence or absence of caveolae (in the case of MLEC). Here, our goal is to probe the variations in cell tension upon hypo-osmotic shock. For the sake of comparison, all the data will be expressed as the relative change of force with respect to the initial tether force f_0 measured in iso-osmotic conditions. The parameter of interest is thus

$$y = \frac{f - f_0}{f_0} \qquad (10.1)$$

10.2.1 Caveolae are Required for Buffering the Tension Surge Due to Hypo-osmotic Shock

As shown in Fig 10.2.1 the measured tether force remained constant upon switch from iso- to hypo-osmotic conditions in wt MLEC, whereas it increased

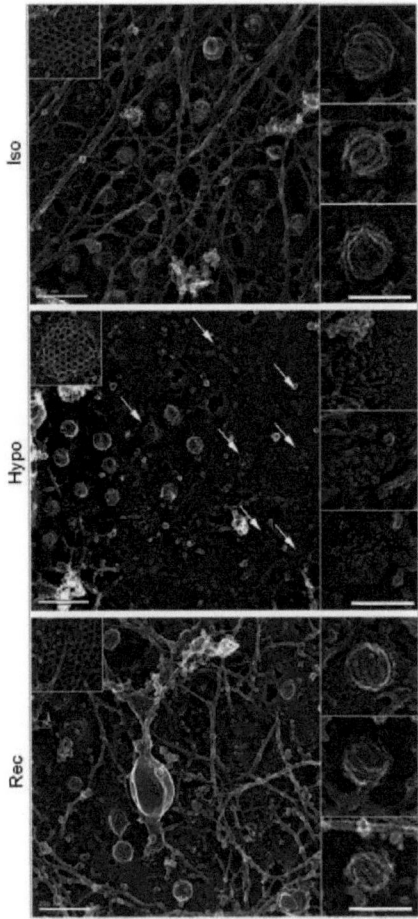

Figure 10.4: Deep-etched EM images of wt MLEC under iso-osmotic (Iso), hypo-osmotic (Hypo) and recovered iso-osmotic (Rec) conditions. Bar = $200nm$. Left insets depict representative images of clathrin-coated pits. Right images depict representative images of caveolae. Bar (insets)= $100nm$.

significantly in cav1$^{-/-}$ cells. This implied that the increase of membrane tension measured in cav1$^{-/-}$ MLEC upon hypo-osmotic shock was buffered in wt MLEC due to the presence of functional caveolae. Accordingly, the expression of functional caveolae by transfection of cav1-EGFP in cav1$^{-/-}$ MLEC re-established the buffering of membrane tension surge in these cells. Furthermore, when cells were pretreated by m-β-cyclodextrin, a treatment known to flatten caveolae through membrane cholesterol depletion ([Rothberg et al., 1992]), the membrane tension increased upon hypo-osmotic shock both in wt and cav1$^{-/-}$ MLEC (Fig 10.2.1 A and B), confirming that the flattening of the invaginated structure of caveolae was required for buffering the surge in membrane tension.

Finally, it was checked whether the buffering effect of caveolae is a peculiarity in MLEC or could be generalized to other cell lines. Thus, the same experiments were conducted with mouse embryonic fibroblasts (MEF) derived from mice lacking or expressing cav1, and with HeLa PFPIG cells. Similar results were obtained with a buffering of membrane tension upon hypo-osmotic shock in wt MEF and HeLa PFPIG, and a membrane tension surge in cav1$^{-/-}$ MEF (Fig 10.2.1).

10.2.2 Clathrin Coated Pits do not Buffer the Membrane Tension

It was tested whether other membrane invaginations such as clathrin-coated pits could play a similar role as caveolae in buffering of membrane tension surge during hypo-osmotic shock. To do that, membrane tension measure-

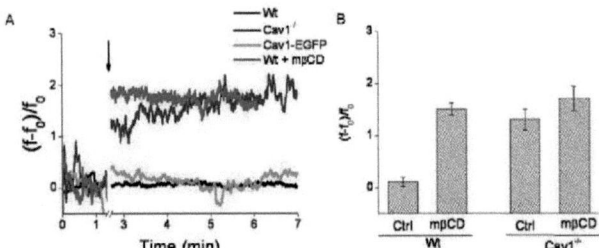

Figure 10.5: (A) Representative force curves for tethers extracted from wt MLEC, cav1$^{-/-}$ MLEC, cav1$^{-/-}$ MLEC transfected with cav1-EGFP, and wt MLEC treated with mβCD exposed to hypo-osmotic shock. After extraction of a tether in iso-osmotic conditions ($0 < t <\sim 1$ minutes), medium is diluted with water until the osmolarity reaches $150 mOsm$ (break from 1.34 to 2.7 minutes indicated by arrow). The membrane tether is maintained at constant length for the whole course of the experiment. The curve represents the relative change of the tether force f with respect to the initial force f_0 measured before hypo-osmotic shock. (B) Relative change of the tether force after hypo-osmotic shock (5 minutes) for wt and cav1$^{-/-}$ MLEC in control conditions ($N = 9$, $p = 0.01502$), and treated with mβCD ($N = 4$, $p = 2E^{-6}$ for wt, and $N = 4$, $p = 0.045$ for cav1$^{-/-}$). f_0 is the resting tether force as measured in iso-osmotic conditions and f is the tether force after hypo-osmotic shock. Data represent mean ± standard errors.

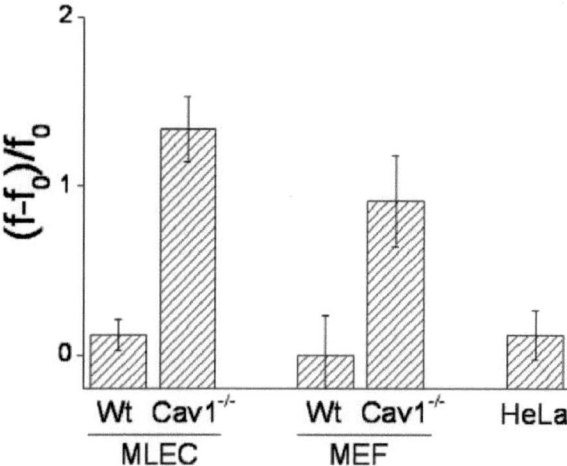

Figure 10.6: Relative change of the tether force after hypo-osmotic shock (5 minutes) for wt ($N = 9$), and cav1$^{-/-}$ MLEC ($N = 9$, $p = 0.01502$), wt ($N = 3$) and cav1$^{-/-}$ MEFs ($N = 4$, $p = 8E^{-4}$) and HeLa PFPIG cells ($N = 4$). f_0 is the resting tether force as measured in iso-osmotic conditions and f is the tether force after hypo-osmotic shock. Data represent mean ± standard errors.

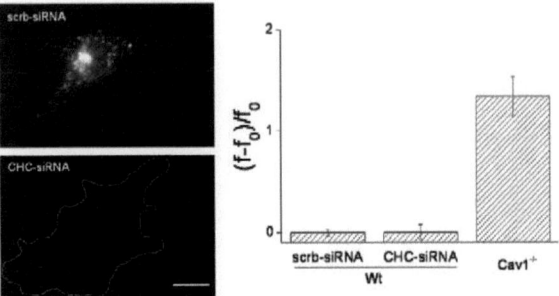

Figure 10.7: (A) Epi fluorescence imaging of transferrin-Alexa594 endocytosis after 5 minutes of incubation in wt MLEC transfected with scrambled siRNA (scrb-siRNA, top) or clathrin heavy chain siRNA (CHC-siRNA, bottom). Cells with no transferrin uptake were considered as ef?ciently transfected with CHC-siRNA, and used for further tether extraction and hypo-shock experiments. Bar = $10 \mu m$. (B) Relative change of the tether force f after hypo-osmotic shock (5 minutes) with respect to the force in iso-osmotic conditions f_0 in wt-MLEC transfected with scrambled siRNA (scrbsiRNA, $N = 7$) or clathrin heavy chain siRNA (CHC-siRNA, $N = 9$), and in cav1$^{-/-}$ MLEC ($N = 10$, $p = 3E^{-4}$) devoid of caveolae, but having clathrin-coated pits. Data represent mean $\pm$ standard errors.

ments were performed on wt MLEC where clathrin was selectively knocked down by RNA interference. Clathrin RNAi led to the inhibition of clathrin-coated pits function as shown by the effcient inhibition of transferrin uptake (Fig 10.2.2 A). However, membrane tension was buffered to the same extent whether clathrin was expressed or knocked down (Fig 10.2.2 B). Contrastingly, cav1$^{-/-}$ MLEC having clathrin-coated pits but no caveolae could not buffer the membrane tension increase under hypo-osmotic shock.

These results ruled out a role for clathrin-coated invaginations in membrane tension buffering and were consistent with our finding that dynasore,

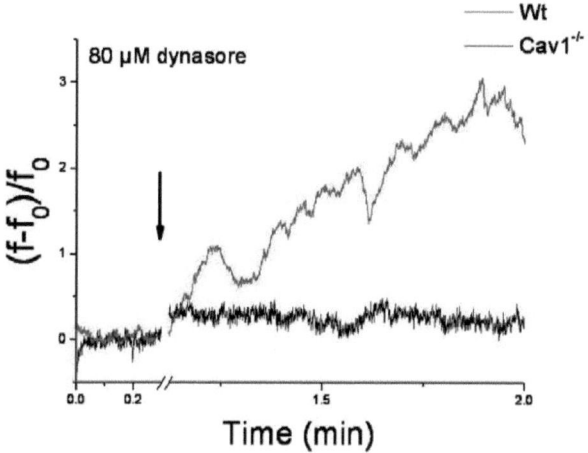

Figure 10.8: Force curves for tethers extracted from wt and cav1$^{-/-}$ MLEC treated with $80\mu M$ dynasore exposed to hypo-osmotic shock. After extraction of a tether in iso-osmotic nconditions ($0 < t < \sim 0.3$ minutes), medium is diluted with water until the osmolarity reaches $150 mOsm$ (break from 0.3 to 0.7 minutes indicated by arrow). The membrane tether is maintained at constant length for the whole course of experiment. The curve represents the relative change of the tether force f with respect to the initial force f_0 measured before hypo-osmotic shock.

which also blocks clathrin-coated pits' endocytosis, did not affect the response of caveolae to hypo-osmotic shock (Fig 10.2.2). Altogether, these results suggest that caveolae behave as primary stress-responsive membrane structures.

10.2.3 Disassembly of Caveolae During Mechanical Stress

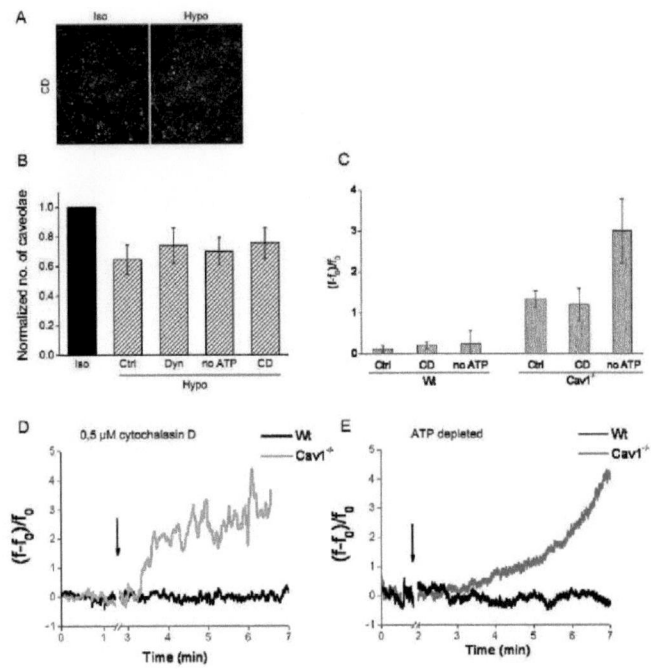

Figure 10.2.3(A) TIRF image of cav1-EGFP in a HeLa cell pretreated with cytochalasin D (CD) at iso-osmotic (Iso) and 5 minutes after switch to hypo-osmotic (Hypo) conditions. Dashed line marks the footprint of the cell. (B) Number of caveolae per cell after hypo-osmotic shock (Hypo) normalized to the number counted in cells before hypo-osmotic shock (Iso) in control (Ctrl; $N = 18$), cytochalasin D (CD; $N = 10$, $p = 2E^{-5}$), latrunculin A (Lat; $N = 21$, $p = 7E^{-11}$), jasplakinolide (Jas, $N = 11$, $p = 6E^{-8}$) treated cells, and ATP depleted cells (no ATP; $N = 10$, $p = 2E^{-5}$). Data represent mean ± standard deviations. (C) Relative change of the tether force f after hypo-osmotic shock (5 minutes) with respect to the force in iso-osmotic conditions f_0 for wt and cav1$^{-/-}$ MLEC for control (Ctrl; $N = 9$ for wt, and for cav1$^{-/-}$ $N = 5$, $p = 0.015$), cytochalasin D treated (CD; $N = 9$ for wt, and for cav1$^{-/-}$ $N = 10$, $p = 3E^{-5}$), and ATP depleted (no ATP; $N = 6$ for wt, and for cav1$^{-/-}$ $N = 5$, $p = 2E^{-4}$) cells. Note that the increase in tether force upon hypo-osmotic shock is significant in each condition where caveolae are absent. Data represent mean ± standard errors. (D), (E) Representative force curves for membrane tethers extracted from wt MLEC (black) and cav1$^{-/-}$ MLEC (grey) treated with (D) cytochalasin D (CD) or (E) ATP depleted (no ATP) and exposed to hypo-osmotic shock. A membrane tether is first extracted in iso-osmotic conditions ($0 < t <\sim$ 1 minutes), and maintained at constant length during the whole course of the experiment. Medium is diluted with water until the osmolarity reaches $150mOsm$ (break from 1 to 2 minutes indicated by arrow). The curve represents the relative change of the tether force f with respect to the initial force f_0. Whereas the magnitude of force increase in cav1$^{-/-}$ MLECs is conserved from one experiment to another, the kinetics of force increase is variable.

Then, it was tested whether the membrane tension buffering effect was actin dependent or an intrinsic property of caveolar invaginations. Cells were thus treated with cytochalasin D to disrupt the cytoskeleton. TIRF imaging however revealed that cytochalasin D did not alter significantly the organization of plasma membrane caveolae nor decreased their number. Additionally, upon hypo-osmotic shock, the disappearance of caveolae was not prevented (Fig 10.2.3 A). In another series of experiments, cells were depleted in ATP. TIRF imaging revealed that the loss of caveolae observed in ATP depleted cells upon hypo-osmotic shock was the same as in untreated cells (Fig 10.2.3 A). As previously described, these treatments decrease the effective cell tension (2.2.3 on page 35). In line with these results, hypo-osmotic shock did not result in an increase of membrane tension in wt MLEC pretreated with cytochalasin D or depleted in ATP. Contrastingly, a drastic membrane tension surge was observed in $cav1^{-/-}$ MLEC subjected to the same treatments (Fig 10.2.3 A). These results indicate that membrane tension buffering is an intrinsic mechanical property of caveolae i.e. it does not require any active cellular machinery.

10.3 Correlation Between the Observed Loss of Caveolae and the Excess of Membrane Area Required to Buffer Membrane Tension

In order to check if the observed loss of caveolae corresponds to the release of additional membrane area required for the membrane tension buffering, the expected membrane area corresponding to the tension increase in caveola free cells will be estimated. To do that, the fact is exploited that the area change of a (reservoir-free) lipid bilayer is directly related to its membrane tension through the Helfrich relation (see 1.1.3):

$$\frac{\sigma}{\sigma_0} = \exp\{\frac{8\pi\kappa}{k_B T}\frac{\Delta A}{A_0}\} \qquad (10.2)$$

where σ_0 is the initial (resting) membrane tension, σ is the membrane tension of the stressed membrane, $k_B T$ is the thermal energy, κ is the bending rigidity ($\sim 40 kT$ as measured on plasma membrane spheres) and A_0 and ΔA are the initial membrane area and change in area, respectively. This relation is an approximation for small area strains and describes the entropic regime, in which bending ripples are progressively smoothened out. We neglect here the elastic regime, in which the membrane exhibits a limited microscopic extensibility, which then leads to lysis. In order to find experimental conditions in which this relation can be reliably used, we selected ATP-depleted or CD-treated cav1$^{-/-}$ MLECs. As abovementioned, the contribution of the cytoskeleton-membrane interaction to the effecive cell tension is minimal (or negligible) upon CD treatment or ATP depletion. The tether force thus directly yields the bilayer tension. Cav1$^{-/-}$ exhibit a significant tether force increase during shock. By measuring f_0 and f, the tether forces in iso-osmotic conditions and after hypo-osmotic shock, we derive σ_0 and σ (since $\sigma = \dfrac{f^2}{8\pi^2\kappa}$ and thus $\Delta A = A_0$, which corresponds to the fractional change in area that cav1$^{-/-}$ MLECs experience during hypo-osmotic shock. Using the numbers obtained for the tether force before and after the hypo-osmotic shock, calculation gives

$$\Delta A/A = 0.28 \pm 0.08\% \qquad (10.3)$$

for ATP-depleted cells (CD treated cells reveal 0.18 ± 0.06% change). Since wt MLECs do not exhibit any tether force increase upon hypo-osmotic shock, our assumption at this stage is that the calculated area change observed with cav1$^{-/-}$ MLECs is assigned to caveola flattening. Now it will be checked, whether this value is close to the effective area released by all the lost cave-olae. Analysis of EM pictures on wt MLEC gave the number of caveolae per μm of membrane section. Calculating the expeected number of caveolae (for example) per $100\mu m^2$ both in iso-osmotic and hypo-osmotic conditions, and then assuming the area per caveola to be $0.02\mu m^2$ ($80nm$ diameter sphere), it is found that caveola flattening leads to 0.38 ± 0.14% change in area. Here, differences of osmolarity between tether force and EM experiments still have to be corrected. Tether forces have been measured at $150mOsm$ while EM was done on samples at $30mOsm$. Using the measured relation between caveola loss and osmolarity (Fig 10.1.1 E), one can expect that caveolae will release a membrane area that corresponds to $0.3\pm0.1\%$ of the total membrane area at $150mOs$. Finally, the estimated area change in $cav1^{-/-}$ MLECs and the fraction of area added by caveolae in wt MLECs are identical within the errors. Although the derivation of these numbers is not direct, the agreement does not seem to us to be coincidental, and supports our proposed mechanism.

Chapter 11

Caveola-mediated Membrane Tension Buffering upon Mechanical Stress: Experiments on Plasma Membrane Spheres

Next, we sought to unambiguously establish that membrane tension buffering is an intrinsic mechanical property of caveolae, i.e. that it does not require any active cellular machinery. To do this, we turned to lipid vesicles, a simplified system to study membrane mechanics. The Bassereau group has a large experience in the reconstitution of membrane proteins into giant unilamellar vesicles (GUVs). The first attempt was the reconstitution of caveolae in GUVs starting from purified cav1 and lipids. However, repetitive failures together with the first publications about cavin1 as a protein required for the formation of functional caveolae ([Hill et al., 2008], [Nabi, 2009], [Hayer et al., 2010]), were the reasons to change our strategy. The idea was to produce caveola containing vesicles directly from cells that are rich in caveolae. Techniques to obtain such vesicles from the native plasma membrane of cells have been developed in the last 40 years ([Scott, 1976], [Holowka and Baird, 1983]). Most of the existing methods imply the use of chemical reagents like formaldehyde or DTT, which can crosslink membrane proteins or drastically alter the intracellular compartments. After our test of these protocols to produce Giant Plasma Membrane Vesicles (GPMV), as described by E. Scott ([Scott, 1976]) and more recently by T. Baumgart and coworkers ([Baumgart et al., 2007]), no caveolae could be observed in the vesicles, but a homogeneous distribution of cav1. In contrast, a more gentle protocol based on the long incubation of cells in phosphate buffer saline (PBS) supplemented with $1.5mM\,Ca^{2+}$ and $1.5mM\,Mg^{2+}$ ([Lingwood et al., 2008], [Kaiser et al., 2009]), allowed us to observe dotted structures that look like caveolae at the membrane of vesicles. These vesicles obtained from detached cellular blebs were named Plasma Membrane Spheres (PMS) by K. Simons. The amount of cav1 found in PMS could be further increased by the use of proteasome inhibitor MG132 ([Lee and Goldberg, 1998]), which inhibits the degradation of proteins by the proteasome.

11.1 Plasma Membrane Spheres Contain Caveolae and Are Devoid of Actin Filaments

11.1.1 Production of PMS from HeLa-PGFPIG

HeLa PFPIG cells expressing cav1-EGFP we re first selected to produce PMS in order to directly assess the presence of caveolar structures in PMS by fluorescence microscopy. After 6 − 10 hours of incubation of cells in PMS-buffer, cav1-EGFP clusters could be observed in cells by epi fluorescence indicating that the treatment did not disassemble caveolae. Similar structures could also be observed in blebs. In a next step, PMS had to be disconnected from the cell in order to manipulate them with a micropipettes and to pull tethers with optical tweezers. In a first attempt, the petri dish containing the cells with the PMS was gently shaken after 6 − 10 hours of incubation, and the liquid was collected in an eppendorf tube, where the vesicles could settle down for 40 minutes, before they were injected into the manipulation chamber placed on the microscope stage. Using this protocol, tether extraction experiments were difficult to perform over longer time periods (i.e. > 1 minute), because of the amount of debris and detached cells which were floating around. Additionally, many PMS were "dirty", i.e. the vesicles contained rigid membrane structures (Fig 11.1.1).

Since only few PMS were needed for the exp eriments, there was no need to obtain a high yield of detached blebs. The protocol was thus changed in such a way that the micropipette was directly used to separate PMS from cells or to pick up free floating PMS. To perform experiments on the Con-OT

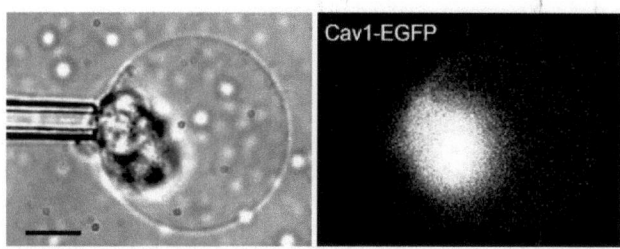

Figure 11.1: Bright field and epi fluorescence image of a "dirty" PMS from a HeLa PFPIG.

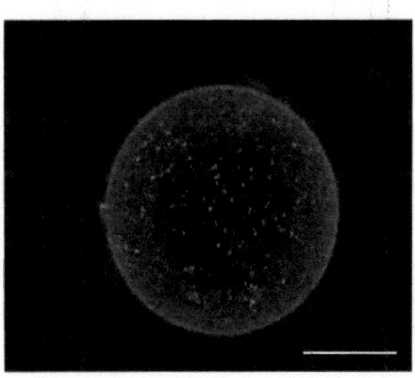

Figure 11.2: Z-stack projection of a "clean" PMS obtained from a HeLa PFPIG. Bar = $10\mu m$.

Figure 11.3: Snapshot of the home made chamber for PMS generation, which could be placed into the cell incubator, and then was mounted on the **Con-OT**

set-up, a new chamber had to be designed based on thick microscopy glass slides, which could be used both for cell treatment in the incubator and for PMS manipulation on the microscope stage (Fig 11.1.1)

11.1.2 Production of PMS from MLEC

PMS could also be generated from wt and cav1$^{-/-}$ MLEC using the same protocol. Although previous reports have shown that PMS were devoid of filamentous actin ([Lingwood et al., 2008]), this finding was double checked by us. To do this, cells were transfected with Lifeact-mCherry, a cytosolic peptide which binds selectively on polymerizing actin. Transfected cells treated for PMS production showed a clear labeling of actin filaments inside the cell, whereas the PMS attached to the cell had only a faint signal of the cytosolic Lifeact-mCherry indicating the absence of filamentous actin (Fig 11.1.2).

In order to test the presence of functional caveolae in PMS, MLECs were

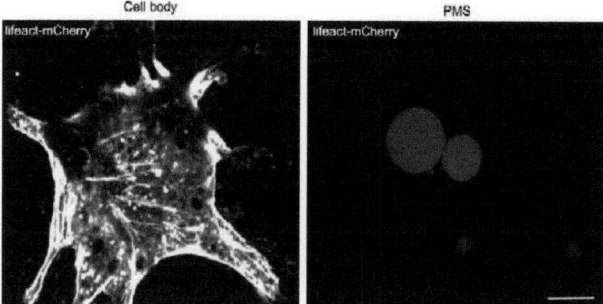

Figure 11.4: Confocal images of a MLEC transfected with Lifeact-mCherry and subjected to PMS production. Lifeact labels the actin filaments in the cell body (left), whereas only a faint signal of cytosolic Lifeact is detected in the PMS sitting on top of the cell (right) indicating the absence of polymerizing actin in PMS

transfected with cav1-EGFP and cavin1-mCherry. Visualization of orange dotted structures in transfected cells indicate co-localization of both proteins, and was taken as the signature of invaginated caveolae ([Hill et al., 2008]). PMS produced from these cells also showed co-localization of cav1-EGFP and cavin1-mCherry at the membrane (Fig 11.1.2 A).

11.2 Micropipette Aspiration of PMS Induces Disassembly of Caveolae

Application of osmotic shock to vesicles is especially delicate because minute shocks are likely to trigger complex changes of the vesicle shape. Instead, micropipette aspiration was used to apply a mechanical stress to PMS. Our goal was to combine simultaneous visualization of caveolae upon aspiration

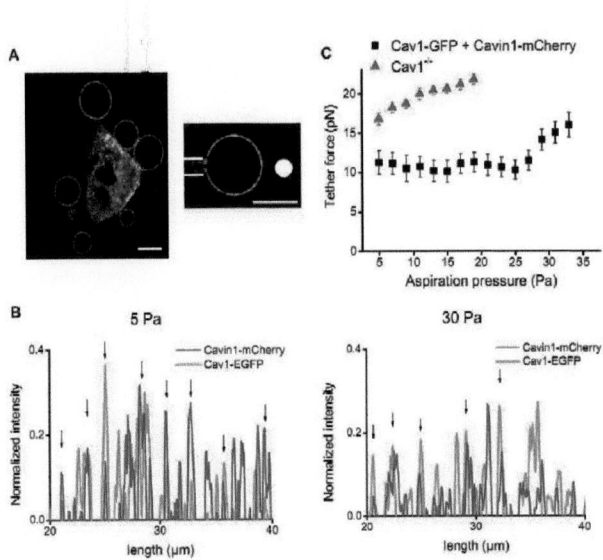

Figure 11.5: (A) Confocal images of a MLEC transfected with cav1-EGFP and cavin1-mCherry after incubation for 6 hours in PMS buffer(left). Cav1 and cavin1 can be found on the membrane of the PMS, even after aspiration by a micropipette (solid white lines) and extraction of a tether by an optically trapped bead (white filled circle)(right). Bar = $10\mu m$. (B) Intensity line scan of the aspirated PMS in (A) at an aspiration pressure of $5Pa$ (left) and $30Pa$ (right). Arrows indicate colocalization of cav1 (green) and cavin1 (red). (C) Plot of the tether force of the PMS from (A) (black) and a PMS obtained from a cav1$^{-/-}$ MLEC (red) in respect to the aspiration pressure. Note that the force of the cav1 containing PMS stays constant until $25Pa$, whereas the force of the PMS lacking cav1 shows a force steadily increasing with the aspiration pressure

and mechanical characterization of the PMS membrane. In a first set of semi-quantitative experiments, PMS were aspirated under increasing suction pressure. Ideally, 3D imaging of PMS should be performed at a given aspiration pressure to obtain a precise estimate of the number of functional caveolae. However, in the absence of cytoskeleton, the diffusion of these structures was too fast to enable confocal z-stack reconstitution. To circumvent this difficulty, the density of caveoale detected at the equator of the vesicle (within a slice of $\sim 0.8/mum$ was checked to be constant in time for a given pressure, and was representative of the overall density of caveolae. Then, quite remarkably, the number of cavin/caveolin co-localization sites within the equatorial slice was observed to decrease upon aspiration (Fig 11.1.2 B). Further, when membrane tether extraction was simultaneously carried out, the tether force was observed to remain constant in the case of PMS containing caveolae while it was increasing for PMS produced from $cav1^{-/-}$ cells. This observation suggests that caveolae disassemble when the PMS is pressurized, as observed in cells submitted to hypo-osmotic shocks. Note that the tether force only remains constant over a limited range of pressure. In the case shown in Fig 11.1.2, the tether force starts to increase above $\Delta P = 25 Pa$. According to the Sens and Turner model ([Sens and Turner, 2005]), which is presented in chapter 4, this tension increase at higher aspiration pressure would suggest that all caveolae are disassembled. Yet, line scans around the PMS perimeter still reveals the presence of cavin/caveolin co-localization sites (Fig 11.1.2 B, right), suggesting either that some co-localization sites are coincidental or that a fraction of caveolae is not able to unfold.

11.2.1 Quantitative Analysis of Micropipette Aspiration of PMS

Micropipette aspiration of PMS from cav1$^{-/-}$ MLEC in combination with tether force measurements allowed to estimate the bending rigidity of PMS from MLECs as $\kappa_{MLEC} = 36 \pm 6\,k_B T$ ($N = 7$, mean± standard error) using the relation 1.2.6 and assuming that the membrane area of cav1$^{-/-}$ PMS is constant, thus, that the membrane tension is set by the aspiration pressure. In the case of PMS from wt MLEC, the height and the length of the observed force plateaus were analyzed. The height of the force plateau is supposed to depend on the mechanical properties of a caveola, whereas the length of the plateau would correspond to the number of caveolae, which contributes to the buffering of membrane tension.

Analysis of the Heigth of the Force Plateau   corresponding to equation 4.0.16 the energy of an invaginated caveola is

$$\bar{\sigma}^{(0)} = S\sigma^{(0)} = (1 - \beta_{bud})^{-1/2}\bar{\gamma} - \bar{\kappa}. \tag{11.1}$$

The height of the force plateaus of 12 vesicles was at $f_{plateau} = 15.7 \pm 0.3\,pN$ (mean ± standard error). With a bending rigidity of PMS of $\sim 40 k_B T$ (as measured for cav1$-/-$ PMS) the membrane tension corresponding to the force plateau is

$$\sigma = \frac{f_{plateau}^2}{8\pi^2 \kappa} = 1.9 \cdot 10^{-5}\,N/m. \tag{11.2}$$

Considering that this is the membrane tension of a budded caveola, one obtains from 4.0.16

$$\bar{\gamma} = \frac{S\sigma^{(0)} + 8\pi\kappa}{\sqrt{1-\beta_{bud}}} \tag{11.3}$$

and with $\beta = 0.9$ and $S = 0.02\mu m^2$

$$\bar{\gamma} = 1.5 \cdot 10^{-19} J. \tag{11.4}$$

This corresponds to a line tension of

$$\bar{\gamma} = \sqrt{\pi S} \cdot \gamma \Rightarrow \gamma = \frac{\bar{\gamma}}{\sqrt{\pi S}} = 6 \cdot 10^{-13} N. \tag{11.5}$$

In summary, the values for the membrane tension of an invaginated caveola annd the corresponding line tension are in a realistic range.

Analysis of the Length of the Force Plateau The additional membrane area necessary for the observed membrane tension buffering is supposed to be provided by the flattening of caveolae. The aspiration pressure until which the force plateau could be observed, was in 12 experiments $P_{crit} = 26 \pm 2 Pa$ (mean $\pm$ standard error), and the starting aspiration pressure was $P_0 = 5Pa$. Considering a lipid membrane vesicle without any membrane reservoir any change in membrane tension is directly related to its areal change by the Helfrich relation (in the entropic regime):

$$\frac{\sigma}{\sigma_{(0)}} = \frac{P}{P_0} = \exp\{\frac{8\pi\kappa}{k_B T}\frac{\Delta A}{A_0}\} \tag{11.6}$$

Using the values obtained from the experiments with PMS from wt MLEC, i.e. $P_{crit} = 26Pa$, $P_0 = 5Pa$, $\kappa = 40k_BT$, one obtains for a vesicle of with radius $R_v = 8\mu m$ an area increase of

$$\Delta A = \frac{4\pi R_v^2 k_B T}{8\pi\kappa} \ln\{\frac{P_{crit}}{P_0}\} = 1.3\mu m^2. \quad (11.7)$$

Thus, the flattening of

$$N_{flat\,cav} = \frac{\Delta A}{S} = 65 \quad (11.8)$$

caveolae (of area $S = 0.02\mu m^2$) would provide the additional membrane area required to buffer the membrane tension. Accordingly to this, the tongue length of three individual PMS from wt-MLEC showing a force plateau upon a similar range of aspiration pressure could be measured and revealed an area increase of $\Delta A = 2.5 \pm 1.2\mu m^2$ (mean ± standard deviation). At the other hand, the number of caveolae in a PMS can be estimated. The confocal image (Fig 11.1.2 A, right) shows an equatorial cut of thickness $e = 0.8\mu m$ and a radius of $R_v = 8\mu m$. For $\Delta P = 5Pa$ $N_{5\,Pa} = 25$ co-localization sites of cav1-EGFP and cavin1-mCherry could be counted, and for $\Delta P = 30Pa$ (at the end of the force plateau) the number of co-localizations was $N_{30\,Pa} = 15$. Note, that N = 0 would be expected at the end of the plateau, meaning that the cavin/caveolin co-localization may be coïncident and does not reflect the existence of functional caveolae.

Nevertheless, consider that the force plateau corresponds for the flattening of $\Delta N_{eq} = 10$ caveolae at the equator of the vesicle with the width $e = 0.8\mu m$ and the radius $R_{eq} = 8\mu m$. Assuming a constant linear caveola

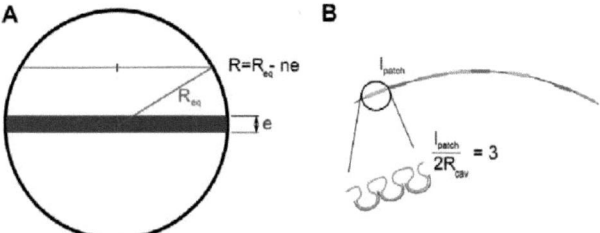

Figure 11.6: (A) Sketch of a PMS depicting the equatorial plane of thickness e scanned with the confocal microscope, and how to estimate from there the number of caveolae in the whole PMS. (B) Sketch of the different section of the line scan of the PMS membrane showing parts with cav1 (green), with cavin1 (red) and with co localization of both (orange). The latter could correspond to multiple caveolae whose number can be estimated in dividing the length of the patch l_{patch} by the diameter of a caveola $2R_{cav}$ (see zoom of the circle).

density, λ, one gets

$$\lambda = \frac{\Delta N_{eq}}{2\pi R_{eq}} = 0.2\mu m^{-1} \tag{11.9}$$

That means for the whole vesicle

$$\Delta N_{ves} = 2\pi R_{eq}\lambda + 2\sum_{n=1}^{R_{eq}/e} 2\pi(R_{eq} - en)\lambda \tag{11.10}$$

$$= \Delta N_{eq}\frac{R_{eq}}{e} = 100 \tag{11.11}$$

caveolae would have flattened upon the increase of aspiration pressure from $5Pa$ to $34Pa$. This number matches the number of caveolae necessary for the membrane buffering calculated in 10.2.8. Another way to estimate the caveola number in the PMS example is to analyze the length of co-localization sites (orange parts) and to put this in relation to the size of a caveola ($R_{cav} = 50nm$), which would provide an upper limit. At $\Delta P = 5Pa$, the 25 patches have a mean length of

$$l^{5Pa}_{patch} = 0.4 \pm 0.2 \mu m$$

and cover a total length of

$$L^{5Pa}_{patch} = 10.3 \mu m.$$

Interestingly, the mean length of the 15 patches at $\Delta P = 30Pa$ increased to

$$l^{30Pa}_{patch} = 0.7 \pm 0.2 \mu m,$$

so that the total length remains at

$$L^{30Pa}_{patch} = 9.9 \mu m.$$

Using equation 10.2.10 one could give an uppper limit of the number of caveolae present in the PMS at $5Pa$ (Fig 11.2.1 B):

$$N^{upper\ limit}_{ves} = \frac{L^{5Pa}_{patch}}{2R_{cav}} \frac{R_{eq}}{e} = 1030. \tag{11.12}$$

Taking into account the number of caveolae at the basal cell membrane detected through TIRF microscopy ($N = \sim 100$), the presence of $50 - 100$ caveolae in a PMS would be realistic, and could explain the observed membrane tension buffering. The high number of co-localization patches could be due to coïncidence and to the mobility of caveolae which can be high in a free membrane without an underlying cytoskeleton. Nevertheless, the observation that the number of co-localization sites decreased, while their mean size increased, indicates that some of the patches corresponded to functional caveolae which disassemble upon mechanical stress.

Chapter 12

Experiments on Muscle Cells The Role of Caveolin-3 Mutations in Muscular Dystrophy

After showing that caveolae play a mechanical role in the response to acute mechanical stress, at short time scales, our focus turned to diseases associated to caveolin mutations, and the question, whether they could originate from the impossibility of altered caveolae to perform this task. The most evident candidates were muscular dystrophies for which many cav3 mutations were reported (see 3.8.2). In collaboration with the laboratory of G. Butler-Browne (Institut de Myologie, Paris), three different muscle cell lines were obtained from biopsies of patients with muscular dystrophies and caveolin-3 mutations:

- cav3-R26Q: associated with LGMD, RMD and HCK

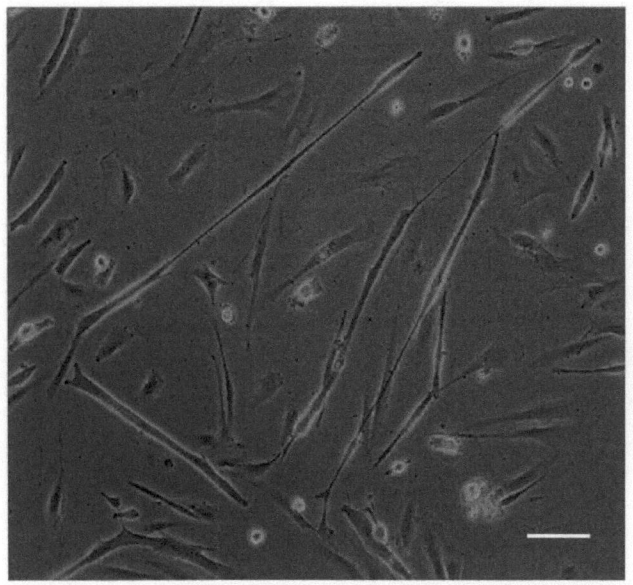

Figure 12.1: Bright field image of wt muscle cells after 7 days in differentiation medium. Myotubes can be identi?ed by their long stretched geometry. Bar =$20\mu m$

- cav3-P28L: associated with HCK

- cav3-A45T: associated with LMGD, RMD

This allowed us to study the case of > 90% reduction of cav3 reported for cav3-R26Q and -A45T, and the less severe 60 – 80% reduction reported for cav3-P28L. Additionally a muscle cell line with wt cav3 was used as a control.

12.1 Tether Force of Differentiated Muscle Cells

Tether extraction on muscle cells revealed that undifferentiated myoblasts could be distinguished from differentiated myotubes (Fig 12) by the tether force. It was in the case of wt cells, $f^{wt,myoblast} = 19 \pm 1.1\,pN$ ($N = 14$) for myoblasts, and $f^{wt,myoblast} = 31 \pm 1.2\,pN$ ($N = 21$) for myotubes (after 7 days of differentiation). Additionally, the tether force of myotubes from the three cell lines with cav3 mutations were similar to the one of wt myotubes, i.e. (Fig 12.1)

- $f^{R26Q,myotube} = 32 \pm 1.1 pN$ ($N = 18$)
- $f^{P28L,myotube} = 31 \pm 1.6 pN$ ($N = 20$)
- $f^{A45T,myotube} = 29 \pm 0.7 pN$ ($N = 18$)

Thus, the effective resting tension of myotubes was unaffected by the presence of functional or mutated caveolae.

12.1.1 Reaction of Myotubes with Cav3-Mutations upon Acute Mechanical Stress

Foll owing the pro cedure describ ed in section 9.2, the response of myotubes to hypo-osmotic shock was tested.

Myotubes are elongated plurinuclear cells ($\sim$ few $100\mu m$ long and $\sim 10\mu m$ wide)(Fig ref11.1). Upon hypo-osmotic shock, a drastic increase of cell volume was observed, irrespective of the cell line (Fig ref11.3). The plot in Fig 12.1.1 shows that the tether force, and thus the membrane tension of wt myotubes is buffered during hypo-osmotic shock. Contrastingly, tether forces

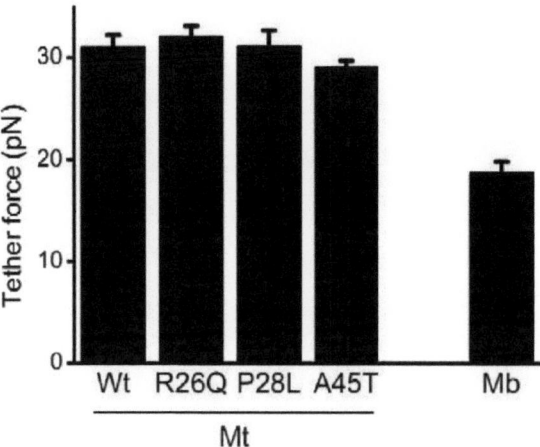

Figure 12.2: Plot showing the tether force of myotubes (Mt) expressing wt cav3 or one of the mutations cav3-R26Q (R26Q), cav3-P28L (P28L), and cav3-A45T (A45T), respectively. The tether force of myotubes is significantly different from undifferentiated wt myoblasts (Mb).

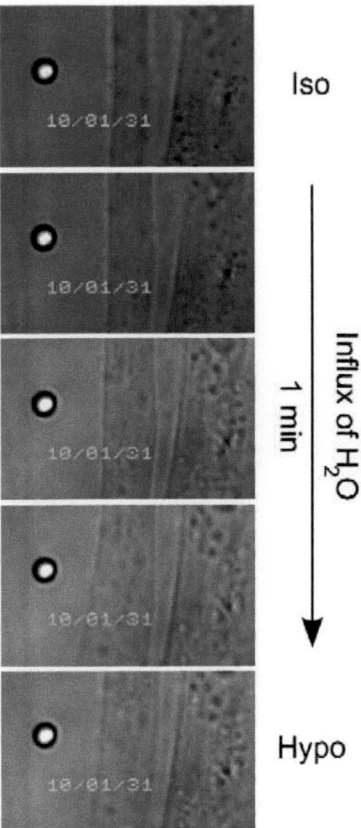

Figure 12.3: Bright field image sequence of a wt myotube during the switch from iso-osmotic (Iso) to hypo-osmotic (Hypo) medium through careful addition of $1ml$ H_2O to $1ml$ cell culture medium during 1 minute. The bead measures $3\mu m$ in diameter.

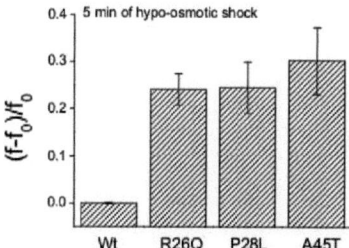

Figure 12.4: Plot showing the relative change of the tether force f after hypo-osmotic shock (5 minutes) in respect to the force in iso-osmotic conditions f_0 in wt, cav3-Q26R (Q26R, $N = 10$, $p = 1E^{-7}$), cav3-P28L (P28L, $N = 13$, $p = 8E^{-7}$), and cav3-A45T (A45T, $N = 7$, $p = 4E^{-7}$) myotubes. Data represent mean ± standard errors.

are observed to increase upon osmotic shock for all the investigated cell lines with cav3-mutations, even though this increase is less drastic than the one reported for MLEC.

12.2 Contracting Myotubes

Occasionally, myotubes started to contract when a tether was extracted with or during addition of water. The contractions lasted for $10 - 20$ seconds and were observed in wt as well as in cells with cav3-mutations (Fig 12.2). This observation has not been exploited further.

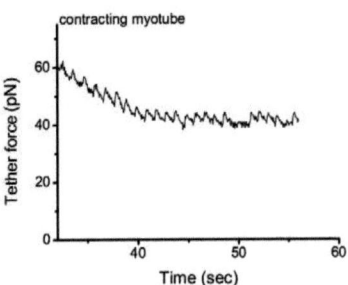

Figure 12.5: Example of a tether force measurement from a contracting myotube. The contractions can be easily identified in the oscillation of the tether force.

Part IV

Discussion

Chapter 13

Caveolae as a Security Device for the Cell Membrane

When acute mechanical stresses induced by hypo-osmotic shock or stretching are applied to cells, the resulting increase of the cell membrane area is expected to lead to an increase of the membrane tension unless additional membrane is released. The main message of the present work is that caveolae may play the role of sacrificial membrane reservoirs to buffer any membrane tension surge at short time scales. In this context, caveolae can be regarded as security devices that prevent the cell membrane from rupturing under acute mechanical stress. This link between mechanical stress, caveola disassembly and membrane tension buffering was deciphered using a combination of several microscopy based techniques. By TIRF-microscopy (mainly performed by my colleague B. Sinha), single caveolae containing cav1-EGFP and co-localizing with cavin1 could be identified at the basal membrane. Application of mechanical stress resulted in a decrease of cav1/cavin1 co-localization, a disappearance of caveolae, and an increase of free cav1, which suggests that caveolae disassemble. This process was independent of ATP and of the actin cytoskeleton. Imaging with EM techniques (performed in collaboration with G. Raposo at the Institut Curie, Paris, and N. Morone at the National Center of Neurology and Psychiatry, Tokyo) allowed similar observations at the ultra structural scale, and showed structures resembling flat caveolae. Finally, experiments of tether extraction with optically trapped beads revealed for the first time that membrane tension surges were buffered in the presence of caveolae, and support the hypothesis that flattening and disassembly of caveolae provide a membrane reservoir that is readily available. In line with the results of TIRF exp eriments, the membrane tension buffering was independent of ATP and actin dynamics, but was impaired when caveolae were disrupted by m-β-cyclodextrin. Accordingly, cells devoid of caveolae were not able to buffer the membrane tension surge upon mechanical stress. Additionally, the production of plasma membrane spheres containing cav1 and cavin1 served as a simplified model system to investigate the influence of caveolae on the mechanical properties of the plasma membrane in the absence of active c ellular mechanisms. As observed in cells, caveolae present at the membrane of PMS buffered the membrane tension upon increasing micropipette aspiration pressure. The study of the mechanical response of caveolae was completed by the observation of caveola reassembly after return from hypotonic to resting conditions. These findings gave rise to the following model of caveola reaction upon me chanical stress which is presented in Fig 13.

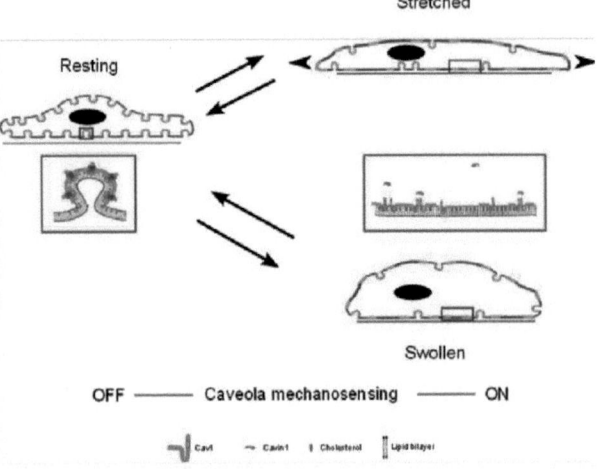

Figure 13.1: Cells Respond to Acute Mechanical Stresses by Rapid Disassembly and Reassembly of Caveolae. In resting conditions, caveolae present at the plasma membrane are mostly invaginated. Magni?cation shows oligomerized cav1 and cavin1 in the caveolar structure. Upon acute mechanical stress (hypo-osmotic shock or stretching), caveolae ?atten out in the plasma membrane to provide additional membrane and buffer membrane tension. Magni?cation shows disassembly and diffusion of cav1 in the plasma membrane and loss ofn interaction between cav1 and cavin1. Return to resting conditions allows the reassembly of the caveolar structure together with cavin1 interaction. This cycle represents the primary cell response to an acute mechanical stress.

13.0.1 Comparison of Experimental Data with the Theoretical Model

Part of the inspiration for the present work originated from the Sens-Turner theoretical model. Before pointing out how it compares to our experimental results, the main assumptions of this theoretical work will be recalled. In this model, invaginated (caveola-like) structures act as a "hidden" membrane reservoir both to set the equilibrium membrane tension and to accommodate tension changes. Briefly, the cell resting state is supposed to be associated with a given number of buds (i.e. functional caveolae), which in turn sets the membrane tension. This claim is straightforward providing that two assumptions are fulfilled: i) the plasma membrane tension is determined by the fraction of flat (or non-invaginated) membrane (hence the term "hidden" for the reservoir of budded structures); ii) at low tension, buds spontaneously form, which means that the budded state is favored in the absence of external source of energy. Then, if tension is increased, buds flatten to release membrane area and buffer the expected tension surge. On the contrary, a drop in membrane tension favors the formation of new buds from the plasma membrane. Therefore, the bold prediction of the model is that the membrane tension remains constant as buds disappear or appear, when the mechanical stress applied to the membrane is respectively tensile or comprssive. In the case of a tensile stress, this buffering process stops when the reservoir of buds is completely depleted.

On the basis of our experimental results, we show that, upon the application of an acute mechanical stress, membrane tension is buffered by caveola disassembly. At first sight, these findings are in line with the Sens-Turner theoretical framework. Yet, closer examination reveals striking differences. From the theoretical point of view, the equilibrium membrane tension is set by the number of buds. However, it is of note that Sens and Turner have only considered a lipid bilayer and neglected any interaction with the cytoskeleton. In this case, the "effective membrane tension" strictly equals the tension of the bilayer (adhesion energy $W_0 = 0$). Conversely, we have long discussed about the influence of the cytoskeleton-membrane adhesion energy W_0 on the effective cell tension that is derived from tether pulling experiments (see section 2.5). First, this term dominates the contribution of the bilayer tension. Second, as seen from our experiments on different cell lines, because it is already difficult to predict the influence of the presence of caveolae on W_0, it is hopeless to think that W_0 could be detectably calibrated by the number of caveolae. In brief, from an experimental point of view, the effective membrane tension seems to be uncorrelated with the number of caveolae. Second, more importantly, the theoretical model was critically tested by applying hyper-osmotic shocks to cells by adding increasing amounts of D-sorbitol to the growth medium used during imaging (data not shown, ob-served by B. Sinha). No significant change in the number of caveolae was observed. This insensitivity to cell deflation is in disagreement with the theoretical model, which means that the conversion between functional caveolae and disassembled caveolae is not a symmetrical process. Along this line, also note that the theoretical model is based on a transition between buds and flat pieces of membrane similar to raft-like domains. Although it cannot be ruled out that, upon hypo-osmotic shock or cell stretching, some of the remaining caveolae are flat (as suggested by deep-etched EM), the fraction of lost caveolae was unambiguously assigned to a molecular disassembly of caveolae. In other words, caveolae are not stable structures in a flat state: they quickly disassemble (within the duration of application of the mechanical stress, $1-2$ minutes) instead of remaining as "rafts". These observations indicate that caveolae do not spontaneously self-assemble from free caveolins available in the membrane, which was a basic hypothesis of the theoretical model (see assumption ii). Finally, although the Turner-Sens model was initially inspiring, it turned out that it could not account for all observations.

Chapter 14

Mechanical Stressand the Role of Caveolae in Signaling

A recent study on cardiac myocytes reported that the presence of caveolae in the plasma membrane leads to a decreased activation of mechanosensitive channels upon application of a mechanical stress ([Kozera et al., 2009]). This observation is consistent with our finding: the activation of mechanosensitive channels is directly dependent on the membrane tension; in consequence, buffering the membrane tension inhibits the activation of the channels. More generally, several proteins and channels implied in signaling pathways were reported to be present in caveolae, and to interact with caveolin proteins [Patel et al., 2008], [Balijepalli and Kamp, 2009]). Some need functional caveolae to be activated upon mechanical stress ([Sedding et al., 2005], [Rizzo et al., 2003], [Park et al., 2000]), whereas others show higher sensitivity to mechanical stress in the absence of caveolae ([Albinsson et al., 2008], [Trouet et al., 1999]). It might be interesting to study, whether these findings can be connected to the mechanical role of caveolae presented in this study following the rationale that

- channels, whose activation depends on membrane tension, are less active in the presence of caveolae

- proteins, which need the interaction with the CSD of single caveolins, would be more activated after caveola disassembly giving rise to higher activation levels in the presence of caveolae ([Head and Insel, 2007])

- proteins, which are concentrated and inactive inside caveolae, would become activated after caveola disassembly (like it is supposed for eNOS)([Patel et al., 2008])

Moreover, it would be important to clarify the role of cavins. On which molecular basis do they associate with caveolin proteins upon caveola formation? Are the cavin proteins (beside cavin1) released from caveolae upon application of a mechanical stress? Do the released cavin proteins activate signaling processes? Cavin1 was first identified as the polymerase 1 and transcript release factor (PTRF) protein ([Jansa et al., 1998]) and as a BF-COL1 (binding factor of a type-1 collagen promoter) binding protein (BBP) ([Hasegawa et al., 2000]). Cavin1 could thus be a link between mechanical stress at the plasma membrane and the regulation of gene transcription in the nucleus. This would imply a transport of cavin1 from the plasma membrane to the nucleus, which still has to be investigated.

Chapter 15

Towards a Better Understanding of Muscular Dystrophies

In the case of muscle cells, the cav3 isoform and its partners are of special interest. Beside the caveolae formed by cav3, this protein is also part of the dystrophin glycoprotein complex (DGC), which is considered to stabilize the plasma membrane ([Lapidos et al., 2004]), and cav3 co-localizes with dysferlin at the plasma membrane, which is important for membrane repair ([Brown and Glover, 2007]). The molecular mechanisms of muscular dystrophies are far from being understood, but one of the reasons could be altered mechanical properties of muscle cells ([Wallace and McNally, 2009]). Many forms of muscular dystrophies occur together with hyper-CKemia, which indicates leaks or ruptures of muscle cell plasma membranes. These dystrophies are as-sociated with mutations of cav3, dysferlin, and members of the DGC, which led to the idea that mutations of these proteins cause an increased rupture of the cell plasma membrane.

In the case of defects of the DGC, the plasma membrane could be more sensitive or more exposed to mechanical stress which results in more rupture events. The DGC connects the extracellular matrix to the cytoskeleton, such that the muscle cells could move "en bloc" together with their extracellular matrix in a muscle fiber. In the case of a lacking link between actin and the extracellular matrix, the movement of muscle cells would be harmonized mechanically, and would pull the extracellular matrix with it. This would lead to an increased friction between the muscle cell plasma membrane and its environment, and lead to more plasma membrane ruptures. This would explain, why the disruption of the actin cytoskeleton evokes the same phenotype as dystrophin lacking cells. The observation of an elevated cav3 expression and caveola density in those cells could indicate a mechanism to compensate the higher membrane stress.

In the case of dysferlin mutations (associated with LGMD-2B) the lack of the rapid membrane repair system established by dysferlin would explain an increased me mbrane rupture rate ([Dowling et al., 2008]).

In the case of caveolinopathies (i.e. disease s associated with caveolin mutations), the lack of a membrane re servoir would be the reason for more frequent membrane ruptures. The present study on myotubes with three different cav3 mutations (R26Q, P28L and A45T) showed a decreased buffering of the membrane tension upon hypo-osmotic shock compared to wt cav3 expressing cells. This could indeed suggest a decreased resistance ot mechanical stress. Because cav3-mutations also alter the dysferlin expression, and its transport to the plasma membrane, further experiments are required to clarify, if the present results are due to the lack of caveolae or to a lack of rapid membrane repair. Monitoring the membrane tension on cells lacking dysferlin, and the study of plasma membrane dysferlin in cells with cav3-mutations upon hypo-osmotic shock would provide information about the action of dysferlin upon mechanical stress. The use of fluorescently tagged cav3 and dystrophin or dysferlin would allow studying their interaction at the plasma membrane using FLIM/FRET. Additional periodic stretch would imitate the situation in real muscles, and would provide information how the muscle adapts to periodic mechanical stress. Monitoring the integrity of the plasma membrane upon acute mechanical stretch could reveal, if the membrane ruptures more easily in cells with defects in the caveola formation and/or if the membrane repair is affected.

Chapter 16

Other Caveolin Related Diseases

Besides the muscular dystrophies other diseases exist which are related to caveolin mutations or dysfunction of caveolae. For example, a young woman was found recently, who lacked cav1 totally and showed amongst others the loss of adipose tissue, increased free tri-cholesterol and lipid levels in the blood (lipo dystrophy), and insulin resistance ([Kim et al., 2008]). Another interesting observation was that people lacking cavin1 lack all caveolae (neither formed by cav1 nor by cav3 in muscle), and thus suffer from muscle dystrophy, lipo dystrophy, insulin resistance ([Hayashi et al., 2009]). But besides these rare and extreme cases, there are mutations of caveolin related to pulmonary hypertension and to cardiac hypertrophy, which are more common ([Patel et al., 2008], [Mercier et al., 2009]). Another big field where the role of caveolin is not understood is cancer. It is thought, that caveolin acts as a tumor suppressor in the early stage of cancer, whereas it is found to be a promoter in metastatic and multiple drugresistant (MDR) cancers ([Goetz et al., 2008]). Since the growth of cancerous tissue implies an increased pressure for the neighbor healthy cells, taking into account mechanical aspects could help to better understand the problem of cancer development. A recent theoretical study of Joanny and coworkers suggested that cells can be transformed to cancer cells by increased pressure, and the study of the role of caveolae in this will be done in our laboratory ([Basan et al., 2009]).

Bibliography

Nabila Aboulaich, Julia P Vainonen, Peter Strå lfors, and Alexander V Vener. Vectorial proteomics reveal targeting, phosphorylation and specific fragmentation of polymerase I and transcript release factor (PTRF) at the surface of caveolae in human adipocytes. *The Biochemical journal*, 383(Pt 2):237–48, 2004. ISSN 1470-8728. doi: 10.1042/BJ20040647.

Sebastian Albinsson, Ina Nordström, Karl Swärd, and Per Hellstrand. Differential dependence of stretch and shear stress signaling on caveolin-1 in the vascular wall. *American journal of physiology. Cell physiology*, 294(1): C271–9, 2008.

Daniel Axelrod. *Total Internal Reflection Fluorescence Microscopy*, volume 89, chapter 7, pages 169–221. Elsevier Inc., 1 edition, 2008.

Ravi C Balijepalli and Timothy J Kamp. Caveolae, ion channels and cardiac arrhythmias. *Progress in biophysics and molecular biology*, 98(2-3):149–60, 2009.

J.-L. Balligand, O. Feron, and C. Dessy. eNOS Activation by Physical Forces: From Short-Term Regulation of Contraction to Chronic Remodeling of Cardiovascular Tissues. *Physiological Reviews*, 80:481–534, 2009.

Dimple Bansal, Katsuya Miyake, Steven S Vogel, Séverine Groh, Chien-Chang Chen, Roger Williamson, Paul L McNeil, and Kevin P Campbell. Defective membrane repair in dysferlin-deficient muscular dystrophy. *Nature*, 423(6936):168–72, May 2003.

Markus Basan, Thomas Risler, Jean-François Joanny, Xavier Sastre-Garau, and Jacques Prost. Homeostatic competition drives tumor growth and metastasis nucleation. *HFSP journal*, 3(4):265–72, August 2009.

Michele Bastiani, Libin Liu, Michelle M Hill, Mark P Jedrychowski, Susan J Nixon, Harriet P Lo, Daniel Abankwa, Robert Luetterforst, Manuel Fernandez-Rojo, Michael R Breen, Steven P Gygi, Jorgen Vinten, Piers J Walser, Kathryn N North, John F Hancock, Paul F Pilch, and Robert G Parton. MURC/Cavin-4 and cavin family members form tissue-specific caveolar complexes. *The Journal of cell biology*, 185(7):1259–73, 2009. ISSN 1540-8140. doi: 10.1083/jcb.200903053.

Manuel Bauer and Lucas Pelkmans. A new paradigm for membrane-organizing and -shaping scaffolds. *FEBS letters*, 580(23):5559–64, October 2006.

Tobias Baumgart, Samuel T. Hess, and Watt W. Webb. Imaging coexisting fluid domains in biomembrane models coupling curvature and the line tension. *Nature*, 425(October):821–824, 2003.

Tobias Baumgart, Adam T Hammond, Prabuddha Sengupta, Samuel T Hess, David A Holowka, Barbara A Baird, and Watt W. Webb. Large-scale fluid/fluid phase separation of proteins and lipids in giant plasma membrane vesicles. *Proceedings of the National Academy of Sciences of the United States of America*, 104(9):3165–70, February 2007.

Andreas Bergdahl and Karl Swärd. Caveolae-associated signalling in smooth muscle. *Canadian journal of physiology and pharmacology*, 82(5):289–99, May 2004.

Regina C. Betz, B G Schoser, Dagmar Kasper, Kenneth Ricker, Alfredo Ramírez, Valentin Stein, Torberg Torbergsen, Y A Lee, Markus M. Nöthen, Thomas F. Wienker, J P Malin, P Propping, André Reis, Wilhelm Mortier, Thomas J. Jentsch, Matthias Vorgerd, and Christian Kubisch. Mutations in CAV3 cause mechanical hyperirritability of skeletal muscle in rippling muscle disease. *Nature genetics*, 28(3):218–9, July 2001.

Timo Betz, Martin Lenz, Jean-François Joanny, and Cécile Sykes. ATP-dependent mechanics of red blood cells. *Proceedings of the National Academy of Sciences of the United States of America*, 106(36):15320–5, September 2009.

Derek J Blake, Andrew Weir, Sarah E Newey, and Kay E Davies. Function and genetics of dystrophin and dystrophin-related proteins in muscle. *Physiological reviews*, 82(2):291–329, April 2002.

Myer Bloom, Evan Evans, and Ole G Mouritsen. Physical properties of the fluid lipid-bilayer component of cell membranes: a perspective. *Quarterly Reviews of Biophysics*, 24(3):293, March 2009.

L. Bo and R. Waugh. Determination of bilayer membrane bending stiffness by tether formation from giant, thin-walled vesicle. *Biophysical Journal*, 55(3):509–517., 1989.

U. Bockelmann, Ph. Thomen, B. Essevaz-Roulet, V. Viasnoff, and F. Heslot. Unzipping DNA with Optical Tweezers: High Sequence Sensitivity and Force Flips. *Biophysical Journal*, 82(3):1537–1553, March 2002.

Eduardo Bonilla, K Fischbeck, and D L Schotland. Freeze-fracture studies of muscle caveolae in human muscular dystrophy. *The American journal of pathology*, 104(2):167–73, August 1981.

E. Botos, J. Klumperman, V. Oorschot, B. Igyártó, A. Magyar, M. Oláh, and A. L. Kiss. Caveolin-1 is transported to multi-vesicular bodies after albumin-induced endocytosis of caveolae in HepG2 cells. *Journal of cellular and molecular medicine*, 12(5A):1632–9, 2007.

B. Bozic, S. Svetina, B. Zeks, and RE Waugh. Role of lamellar membrane structure in tether formation from bilayer vesicles. *Biophysical journal*, 61(4):963–973, 1992.

A. Bretscher. Microfilament structure and function in the cortical cytoskeleton. *Annual review of cell biology*, 7(1):337–374, 1991.

Robert H Brown and Louise Glover. Dysferlin in membrane trafficking and patch repair. *Traffic (Copenhagen, Denmark)*, 8(7):785–94, 2007.

Michael R. Bubb, Adrian M. J. Senderowicz, Edward A. Sausville, Kimberly L. K. Duncan, and Edward D. Korn. Jasplakinolide, a Cytotoxic Natural Product, Induces Actin Polymerization and Competitively Inhibits the

Binding of Phalloidin to F-actin. *The Journal of biological chemistry*, 269 (May 27):14869–14871, 1994.

Roger Burgener, Marlene Wolf, T Ganz, and Marco Baggiolini. Purification and characterization of a major phosphatidylserine-binding phosphoprotein from human platelets. *The Biochemical journal*, 269(3):729–34, August 1990.

W.S. Bush, G. Ihrke, J.M. Robinson, and A.K. Kenworthy. Antibody-specific detection of caveolin-1 in subapical compartments of MDCK cells. *Histochemistry and cell biology*, 126(1):27–34, 2006.

Katharine M.D. Bushby. The limb-girdle muscular dystrophies-multiple genes, multiple mechanisms. *Human molecular genetics*, 8(10):1875–82, January 1999.

Xavier Buton, Gil Morrot, Pierre Fellmann, and Michel Seigneuret. Ultrafast Glycerophospholipid-selective Transbilayer Motion Mediated by a Protein in the Endoplasmic Reticulum Membrane*. *Biochemistry*, 271(12):6651–6657, 1996.

Felix Campelo, Harvey T McMahon, and Michael M Kozlov. The hydrophobic insertion mechanism of membrane curvature generation by proteins. *Biophysical journal*, 95(5):2325–39, September 2008.

P. B. Canham. The minimum energy of bending as a possible explanation of the biconcave shape of the human red blood cell. *Journal of theoretical biology*, 26(1):61–81, January 1970.

A. J. Carozzi, S. Roy, I. C. Morrow, A. Pol, B. Wyse, J. Clyde-smith, I. A. Prior, S. J. Nixon, J. F. Hancock, and R. G. Parton. Inhibition of lipid raft-dependent signaling by a dystrophy-associated mutant of caveolin-3. *J. Biol. Chem*, 277:17944–17949, 2002.

Y. Chen and L. C. Norkin. Extracellular simian virus 40 transmits a signal that promotes virus enclosure within caveolae. *Experimental cell research*, 246(1):83–90, January 1999.

Nam-Hai Chuat, Hans Christensen, Marie-France Carlier, Irina Perelroizen, and Dominique Didry. Role of nucleotide exchange and hydrolysis in the function of profilin and actin assembly. *The Journal of Biological Chemistry,*, 271:1230212309, 1996.

J. A. Cooper. Effects of cytochalasin and phalloidin on actin. *J. Cell Biol.*, 105:1473–1478, 1987.

D. D. W. Cornelison. Context matters: in vivo and in vitro influences on muscle satellite cell activity. *Journal of cellular biochemistry*, 105(3):663–9, October 2008.

G. Cossu, S. Tajbakhsh, and M. Buckingham. How is myogenesis initiated in the embryo? *Trends in genetics : TIG*, 12(6):218–23, June 1996.

Harold Couchoux, Bruno Allard, Claude Legrand, Vincent Jacquemond, and Christine Berthier. Loss of caveolin-3 induced by the dystrophy-associated P104L mutation impairs L-type calcium channel function in mouse skeletal muscle cells. *The Journal of physiology*, 580(Pt.3):745–54, May 2007.

Damien Cuvelier. *Adhésion cellulaire et tubes de membrnae: Quelques aspects dynamiques, mécaniques et rhéologiques.* PhD thesis, 2005.

Damien Cuvelier, Imre Derényi, Patricia Bassereau, Pierre Nassoy, and Imre Dere. Coalescence of membrane tethers: experiments, theory, and applications. *Biophysical journal*, 88(4):2714–26, 2005.

J. Dai, M. P. Sheetz, X. Wan, and C. E. Morris. Membrane tension in swelling and shrinking molluscan neurons. *J. Neurosci*, 18:6681–6692, 1998.

Jianwu Dai, H P Ting-Beall, and Michael P Sheetz. The secretion-coupled endocytosis correlates with membrane tension changes in RBL 2H3 cells. *The Journal of general physiology*, 110(1):1–10, July 1997.

Mariella D'Alessandro, David Russell, Susan M. Morley, Anthony M. Davies, and E. Birgitte Lane. Keratin mutations of epidermolysis bullosa simplex alter the kinetics of stress response to osmotic shock. *Journal of cell science*, 115(Pt 22):4341–51, November 2002.

Isin Dalkilic and Louis M Kunkel. Muscular dystrophies: genes to pathogenesis. *Current opinion in genetics and development*, 13(3):231–8, June 2003.

Gawiyou Danialou, Alain S Comtois, Roy Dudley, George Karpati, Geneviève Vincent, C Des Rosiers, and Basil J Petrof. Dystrophin-deficient cardiomyocytes are abnormally vulnerable to mechanical stress-induced contractile failure and injury. *The FASEB journal : official publication of the Federation of American Societies for Experimental Biology*, 15(9):1655–7, July 2001.

E. E. Daniel and W. J. Cho. Caveolae and calcium handling, a review and a hypothesis. 2006.

I. Derényi, F. Julicher, and J. Prost. Formation and interaction of membrane tubes. *Physical review letters*, 88(23):238101, 2002.

James J. Dowling, Elizabeth M. Gibbs, and Eva L. Feldman. Membrane traffic and muscle: lessons from human disease. *Traffic (Copenhagen, Denmark)*, 9(7):1035–43, 2008.

J. L. Drury and M. Dembo. Hydrodynamics of micropipette aspiration. *Biophysical journal*, 76(1 Pt 1):110–28, January 1999.

Angela F. Dulhunty and Clara Franzini-Armstrong. The Relative Contributions of the Folds and Caveolae to the Surface Membrane of Frog Skeletal Muscle Fibres at Different Sarcomere Lengths. *Journal of Physiology*, 250: 513–539, 1975.

Madeleine Durbeej and Kevin P Campbell. Muscular dystrophies involving the dystrophin-glycoprotein complex: an overview of current mouse models. *Current opinion in genetics and development*, 12(3):349–61, June 2002.

Asier Echarri, Olivia Muriel, and Miguel A Del Pozo. Intracellular trafficking of raft/caveolae domains: insights from integrin signaling. *Seminars in cell and developmental biology*, 18(5):627–37, October 2007.

Richard M. Epand, Brian G. Sayer, and Raquel F. Epand. Caveolin scaffolding region and cholesterol-rich domains in membranes. *Journal of molecular biology*, 345(2):339–50, 2005.

E. Evans and W. Rawicz. Entropy-driven tension and bending elasticity in condensed-fluid membranes. *Physical Review Letters*, 64:2094–2097, 1990a.

E. Evans and W. Rawicz. Entropy-driven tension and bending elasticity in condensed-fluid membranes. *Physical Review Letters*, 64(17):2094–2097, April 1990b.

E. Evans and A. Yeung. Hidden dynamics in rapid changes of bilayer shape. *Chemistry and Physics of Lipids*, 73(1-2):39–56, September 1994.

Evan Evans, Volkmar Heinrich, Florian Ludwig, and Wieslawa Rawicz. Dynamic Tension Spectroscopy and Strength of Biomembranes. *Biophysical Journal*, 85(4):2342–2350, October 2003.

M. Faris, D. Lacoste, J. Pécréaux, J.-F. Joanny, J. Prost, and P. Bassereau. Membrane Tension Lowering Induced by Protein Activity. *Physical Review Letters*, 102(3), January 2009.

J.F. Faucon, M. D. Mitov, P. Méléard, I. Bivas, and P. Bothorel. Bending elasticity and thermal fluctuations of lipid membranes. Theoretical and experimental requirements. *Journal de Physique*, 50(17):2389–2414, 1989.

I. Fernandez, Y. Ying, J. Albanesi, and R. G. W. Anderson. Mechanism of caveolin filament assembly. *Proc. Natl. Acad. Sci. U.SA.*, 99:11193–111198, 2002.

Daniel A. Fletcher and R. Dyche Mullins. Cell mechanics and the cytoskeleton. *Nature*, 463(7280):485–92, January 2010.

J.-B. Fournier, A. Ajdari, and L. Peliti. Effective-Area Elasticity and Tension of Micromanipulated Membranes. *Physical Review Letters*, 86(21):4970–4973, May 2001.

Nicholas Franki, Guohua Ding, Yang Gao, and Richard M. Hays. Effect of cytochalasin D on the actin cytoskeleton of the toad bladder epithelial cell. *American Physiological Society*, pages C995–1000, 1992.

K. Fricke, K. Wirthensohn, R. Laxhuber, and E. Sackmann. Flicker spectroscopy of erythrocytes. A sensitive method to study subtle changes of membrane bending stiffness. *European biophysics journal : EBJ*, 14(2): 67–81, January 1986.

L. Miya Fujimoto, Robyn Roth, John E. Heuser, and Sandra L. Schmid. Actin Assembly Plays a Variable, but not Obligatory Role in Receptor-Mediated Endocytosis in Mammalian Cells. *Traffic*, (6):161–171, 2000.

T. Fujimoto. Calcium pump of the plasma membrane is localized in caveolae. *The Journal of cell biology*, 120(5):1147–57, March 1993.

F. Galbiati, B. Razani, and M. P. Lisanti. Caveolae and caveolin-3 in muscular dystrophy. *Trends Mol. Med*, 7:435–441, 2001.

Elisabetta Gazzerro, Federica Sotgia, Claudio Bruno, Michael P Lisanti, and Carlo Minetti. Caveolinopathies: from the biology of caveolin-3 to human diseases. *European Journal of Human Genetics*, 18(April):137–145, 2010.

Benjamin Geiger, Alexander Bershadsky, Roumen Pankov, and K M Yamada. Transmembrane crosstalk between the extracellular matrix-cytoskeleton crosstalk. *Nature reviews. Molecular Cell Biology*, 2(11):793–805, November 2001.

M. Glogauer, P. Arora, G. Yao, I. Sokholov, J. Ferrier, and C. A. McCulloch. Calcium ions and tyrosine phosphorylation interact coordinately with actin to regulate cytoprotective responses to stretching. *Journal of cell science*, 110 (Pt 1:11–21, January 1997.

Jacky G. Goetz, Patrick Lajoie, Sam M. Wiseman, and Ivan R. Nabi. Caveolin-1 in tumor progression: the good, the bad and the ugly. *Cancer metastasis reviews*, 27(4):715–35, 2008.

Araceli Grande-García and Miguel a del Pozo. Caveolin-1 in cell polarization and directional migration. *European journal of cell biology*, 87(8-9):641–7, September 2008.

I. Grummt. Regulation of mammalian ribosomal gene transcription by RNA polymerase I. *Prog. Nucleic Acid Res. Mol. Biol.*, 62:109–154, 1999.

S. Gustincich and C. Schneider. Serum deprivation response gene is induced by serum starvation but not by contact inhibition. *Cell growth and differentiation : the molecular biology journal of the American Association for Cancer Research*, 4(9):753–60, September 1993.

Andrew J. Halayko, Blanca Camoretti-Mercado, Sean M. Forsythe, Joaquim E. Vieira, Richard W. Mitchell, Mark E. Wylam, Marc B. Hershenson, and Julian Solway. Divergent differentiation paths in airway smooth muscle culture: induction of functionally contractile myocytes. *Am J Physiol Lung Cell Mol Physiol*, 276:197–206, 1999.

John F Hancock. Lipid rafts: contentious only from simplistic standpoints. *Nature reviews. Molecular Cell Biology*, 7(6):456–62, June 2006.

Carsten G Hansen and Ben J Nichols. Exploring the caves: cavins, caveolins and caveolae. *Trends in cell biology*, pages 1–10, 2010. ISSN 1879-3088. doi: 10.1016/j.tcb.2010.01.005.

Carsten G Hansen, Nicholas A Bright, Gillian Howard, and Benjamin J Nichols. SDPR induces membrane curvature and functions in the formation of caveolae. *Nature cell biology*, 11(7):807–14, July 2009.

T. Hasegawa, A. Takeuchi, O. Miyaishi, H. Xiao, J. Mao, and K. Isobe. PTRF (polymerase I and transcript-release factor) is tissue-specific and interacts with the BFCOL1 (binding factor of a type-I collagen promoter) zinc-finger transcription factor which binds to the two mouse type-I collagen gene promoters. *The Biochemical journal*, 347 Pt 1:55–9, April 2000.

Yukiko K. Hayashi, Young-Eun Park, Chie Matsuda, Naomi Hino-Fukuyo, Megumu Ogawa, Kanako Goto, Kayo Tominaga, Satomi Mitsuhashi, Ikuya Nonaka, Kazuhiro Haginoya, Hisashi Sugano, and Ichizo Nishino. Human PTRF mutations cause secondary deficiency of caveolins resulting in muscular dystrophy with generalized lipodystrophy. *The Journal of clinical investigation*, 119(9):2623–33, 2009.

Arnold Hayer, Miriam Stoeber, Christin Bissig, and Ari Helenius. Biogenesis of caveolae: stepwise assembly of large caveolin and cavin complexes. *Traffic (Copenhagen, Denmark)*, 11(3):361–82, March 2010.

Brian P. Head and Paul A. Insel. Do caveolins regulate cells by actions outside of caveolae? *Trends in cell biology*, 17(2):51–7, 2007.

V. Heinrich and R. Waugh. A piconewton force transducer and its application to measurement of the bending stiffness of phospholipid membranes. *Ann Biomed Eng*, 24:595–605, 1996.

Volkmar Heinrich, Andrew Leung, and Evan Evans. Nano- to microscale dynamics of P-selectin detachment from leukocyte interfaces. II. Tether flow terminated by P-selectin dissociation from PSGL-1. *Biophysical journal*, 88(3):2299–308, March 2005.

W. Helfrich. Elastic properties of lipid bilayers: theory and possible experiments. *Zeitschrift für Naturforschung. Teil C: Biochemie, Biophysik, Biologie, Virologie*, 28(11):693–703, 1973.

W. Helfrich and R. M. Servuss. Undulations, steric interaction and cohesion of fluid membranes. *Il Nuovo Cimento D*, 3(1):137–151, January 1984.

S. Hénon, G. Lenormand, A. Richert, and F. Gallet. A new determination of the shear modulus of the human erythrocyte membrane using optical tweezers. *Biophysical journal*, 76(2):1145–51., 1999.

Marc Herant, Volkmar Heinrich, and Micah Dembo. Mechanics of neutrophil phagocytosis: behavior of the cortical tension. *Journal of cell science*, 118 (Pt 9):1789–97, May 2005.

Delia J Hernández-Deviez, Sally Martin, Steven H Laval, Harriet P Lo, Sandra T Cooper, Kathryn N North, Kate Bushby, and Robert G Parton. Aberrant dysferlin trafficking in cells lacking caveolin or expressing dystrophy mutants of caveolin-3. *Human molecular genetics*, 15(1):129–42, 2006.

Ralf Herrmann, Volker Straub, Martina Blank, Christian Kutzick, Nicola Franke, Eva Neuen Jacob, H G Lenard, Stephan Kröger, and Thomas Voit. Dissociation of the dystroglycan complex in caveolin-3-deficient limb girdle muscular dystrophy. *Human molecular genetics*, 9(15):2335–40, September 2000.

Michelle M. Hill, Michele Bastiani, Robert Luetterforst, Matthew Kirkham, Annika Kirkham, Susan J. Nixon, Piers Walser, Daniel Abankwa, Viola M. J. Oorschot, Sally Martin, John F. Hancock, and Robert G. Parton. PTRF-Cavin, a conserved cytoplasmic protein required for caveola formation and function. *Cell*, 132(1):113–24, January 2008.

Lars Hinrichsen, Anika Meyerholz, Stephanie Groos, and Ernst J Ungewickell. Bending a membrane: how clathrin affects budding. *Proceedings of the National Academy of Sciences of the United States of America*, 103(23): 8715–20, June 2006.

F. M. Hochmuth, J. Y. Shao, Jianwu Dai, and Michael P. Sheetz. Deformation and flow of membrane into tethers extracted from neuronal growth cones. *Biophysical journal*, 70(1):358–69, January 1996.

R. Hochmuth, H. WILES, and E. Evans. Extensionnal flow of erythrocyte-membrane from cell body to elastic tether. *Biophysical journal*, 39(1): 83–89, 1982.

R. M. Hochmuth. Micropipette aspiration of living cells. *Journal of Biomechanics*, 133:305–323, 2000.

R. M. Hochmuth, N. Mohandas, and P. L. Blackshear, JR. Measurement of the Elastic Modulus for Red Cell Membrane Using a Fluid Mechanical Technique. *Biophysical Journal*, 13:747–762, 1973.

David Holowka and Barbara Baird. Structural Studies on the Membrane-Bound Immunoglobulin E-Receptor Complex. 1. Characterization of Large Plasma Membrane Vesicles from Rat Basophilic Leukemia Cells and Insertion of Amphipathic Fluorescent Probes? *Biochemistry*, (1979):3466–3474, 1983.

Andreas Holzinger and Ursula Meindl. Jasplakinolide, a Novel Actin Targeting Peptide, Inhibits Cell Growth and Induces Actin Filament Polymerization in the Green Alga. *Cell Motility and the Cytoskeleton*, 38(May): 365–372, 1997.

Yasushi Izumi, S I Hirai, Yoko Tamai, A Fujise-Matsuoka, Yoshifumi Nishimura, and Shigeo Ohno. A protein kinase Cdelta-binding protein SRBC whose expression is induced by serum starvation. *The Journal of biological chemistry*, 272(11):7381–9, March 1997.

P. A. Janmey and P. K. J. Kinnunen. Biophysical properties of lipids and dynamic membranes. *Trends in cell biology*, 16(10):538–46, October 2006.

P. Jansa, S. W. Mason, U. Hoffmann-Rohrer, and I. Grummt. Cloning and functional characterization of PTRF, a novel protein which induces dissociation of paused ternary transcription complexes. *The EMBO journal*, 17 (10):2855–64, May 1998.

Hermann-Josef Kaiser, Daniel Lingwood, Ilya Levental, Julio L Sampaio, Lucie Kalvodova, Lawrence Rajendran, and Kai Simons. Order of lipid phases in model and plasma membranes. *Proceedings of the National Academy of Sciences of the United States of America*, 106(39):16645–50, September 2009.

C. A. Kim, Marc Delépine, Emilie Boutet, Haquima El Mourabit, Soazig Le Lay, Muriel Meier, Mona Nemani, Etienne Bridel, Claudia C Leite, Debora R Bertola, Robert K Semple, Stephen O'Rahilly, Isabelle Dugail, Jacqueline Capeau, Mark Lathrop, and Jocelyne Magré. Association of a homozygous nonsense caveolin-1 mutation with Berardinelli-Seip congenital lipodystrophy. *The Journal of clinical endocrinology and metabolism*, 93(4):1129–34, April 2008.

Edgar E. Kooijman, Vladimir Chupin, Nola L. Fuller, Michael M. Kozlov, Ben de Kruijff, Koert N. J. Burger, and Peter R. Rand. Spontaneous curvature of phosphatidic acid and lysophosphatidic acid. *Biochemistry*, 44(6):2097–102, February 2005.

Gerbrand Koster, Martijn VanDuijn, Bas Hofs, and Marileen Dogterom. Membrane tube formation from giant vesicles by dynamic association of motor proteins. *Proceedings of the National Academy of Sciences of the United States of America*, 100(26):15583-8, December 2003.

Gerbrand Koster, Angelo Cacciuto, Imre Derényi, Daan Frenkel, and Marileen Dogterom. Force Barriers for Membrane Tube Formation. *Physical Review Letters*, 94(6):16–19, February 2005.

Lukasz Kozera, Ed White, and Sarah Calaghan. Caveolae act as membrane reserves which limit mechanosensitive I(Cl,swell) channel activation during swelling in the rat ventricular myocyte. *PloS one*, 4(12):e8312, January 2009.

Michael Kristensen, Martin Krø yer Rasmussen, and Carsten Juel. Na(+)-K (+) pump location and translocation during muscle contraction in rat skeletal muscle. *Pflügers Archiv : European journal of physiology*, 456(5): 979–89, 2008.

T. V. Kurzchalia, P. Dupree, R. G. Parton, R. Kellner, H. Virta, M. Lehnert, and K. Simons. VIP21, a 21-kD membrane protein is an integral component of trans-Golgi-network-derived transport vesicles. *The Journal of cell biology*, 118(5):1003–14, September 1992.

R. Kwok and E. Evans. Thermoelasticity of large lecithin bilayer vesicles. *Biophysical Journal*, 35(3):637–652, 1981.

Karen A. Lapidos, Rahul Kakkar, and Elizabeth M McNally. The dystrophin glycoprotein complex: signaling strength and integrity for the sarcolemma. *Circulation research*, 94(8):1023–31, 2004.

S. H. Laval and K. M. D. Bushby. Limb-girdle muscular dystrophies–from genetics to molecular pathology. *Neuropathology and applied neurobiology*, 30(2):91–105, April 2004.

Daniel J Leary and Sui Huang. Regulation of ribosome biogenesis within the nucleolus. *FEBS Letters*, 509:145–150, 2001.

D. H. Lee and A. L. Goldberg. Proteasome inhibitors: valuable new tools for cell biologists. *Trends in cell biology*, 8(10):397–403, October 1998.

Niall J. Lennon, Alvin Kho, Brian J. Bacskai, Sarah L. Perlmutter, Bradley T. Hyman, and Robert H. Brown. Dysferlin interacts with annexins A1 and A2 and mediates sarcolemmal wound-healing. *The Journal of biological chemistry*, 278(50):50466–73, December 2003.

Shengwen Li, Kenneth S. Song, and Michael P. Lisanti. Expression and Characterization of Recombinant Caveolin. *The Journal of biological chemistry*, 271(1):568–573, 1996.

Daniel Lingwood, Jonas Ries, Petra Schwille, and Kai Simons. Plasma membranes are poised for activation of raft phase coalescence at physiological temperature. *Proceedings of the National Academy of Sciences of the United States of America*, 105(29):10005–10, July 2008.

R. Lipowsky and E. Sackmann. *Structure and Dynamics of Membranes: from Cells to Vesicles*, volume 1A. Elsevier sciences B.V., 1995.

R. Lipowsky, H. Döbereiner, C. Hiergeist, and V. Indrani. Membrane curvature induced by polymers and colloids. *Physica A: Statistical and Theoretical Physics*, 249(1-4):536–543, February 1998.

Libin Liu and Paul F Pilch. A critical role of cavin (polymerase I and transcript release factor) in caveolae formation and organization. *The Journal of biological chemistry*, 283(7):4314–22, February 2008.

Libin Liu, Dennis Brown, Mary McKee, Nathan K Lebrasseur, Dan Yang, Kenneth H Albrecht, Katya Ravid, and Paul F Pilch. Deletion of Cavin/PTRF causes global loss of caveolae, dyslipidemia, and glucose intolerance. *Cell metabolism*, 8(4):310–7, 2008.

Ange Maguy, Terence E Hebert, and Stanley Nattel. Involvement of lipid rafts and caveolae in cardiac ion channel function. *Current Opinion in Cell Biology*, 69:798 – 807, 2006.

A. K. Malik, P. E. Monahan, D. L. Allen, B. G. Chen, R. J. Samulski, and K. Kurachi. Kinetics of recombinant adeno-associated virus-mediated gene transfer. *Journal of virology*, 74(8):3555–65, April 2000.

J. B. Manneville, P. Bassereau, S. Ramaswamy, and J. Prost. Active membrane fluctuations studied by micropipet aspiration. *Physical review. E, Statistical, nonlinear, and soft matter physics*, 64(2 Pt 1):021908, August 2001.

D Marsh. Elastic curvature constants of lipid monolayers and bilayers. *Chemistry and Physics of Lipids*, 144:146–159, 2006.

Kerrie-Ann McMahon, Hubert Zajicek, Wei-Ping Li, Michael J Peyton,

John D Minna, V James Hernandez, Katherine Luby-Phelps, and Richard G W Anderson. SRBC/cavin-3 is a caveolin adapter protein that regulates caveolae function. *The EMBO journal*, 28(8):1001–15, 2009.

P. Meleard, C. Gerbeaud, P. Bardusco, N. Jeandaine, M. D. Mitov, and L. Fernandez-Puente. Mechanical properties of model membranes studied from shape transformations of giant vesicles. *Biochimie*, 80(5-6):401–413, May 1998.

Isabelle Mercier, Jean-Francois Jasmin, Stephanos Pavlides, Carlo Minetti, Neal Flomenberg, Richard G Pestell, Philippe G Frank, Federica Sotgia, and Michael P Lisanti. Clinical and translational implications of the caveolin gene family: lessons from mouse models and human genetic disorders. *Laboratory investigation; a journal of technical methods and pathology*, 89(6):614–23, 2009.

L. Merlini, I. Carbone, C. Capanni, P. Sabatelli, S. Tortorelli, F. Sotgia, M. P. Lisanti, C. Bruno, and C. Minetti. Familial isolated hyperCKaemia associated with a new mutation in the caveolin-3 (CAV-3) gene. *Journal of neurology, neurosurgery, and psychiatry*, 73(1):65–7, July 2002.

L. R. Mills and C. E. Morris. Neuronal plasma membrane dynamics evoked by osmomechanical perturbations. *The Journal of membrane biology*, 166(3):223–35, December 1998.

N. Mohandas and E. Evans. Mechanical properties of the red cell membrane in relation to molecular structure and genetic defects. *Annual Reviews of Biophysics Biomolecular Structures*, 23:787–818, 1994.

Solange Monier, Robert G. Parton, Frank Vogel, Joachim Behlke, Annemarie Henske, and Teymuras V Kurzchalia. VIP21-Cavolin, a membrane protein constituent of the caveolar coat, oligomerizes in vivo and in vitro. *Molecular Biology of the Cell*, 6:911–927, 1995.

Nobuhiro Morone. Freeze-etch electron tomography for the plasma membrane interface. *Methods in molecular biology (Clifton, N.J.)*, 657(1):275–86, January 2010.

C.E. Morris and U. Homann. Cell Surface Area Regulation and Membrane Tension. *Membrane Biology*, 179:79–102, 2001.

Takahisa Murata, Michelle I Lin, Radu V Stan, Phillip Michael Bauer, Jun Yu, and William C Sessa. Genetic evidence supporting caveolae microdomain regulation of calcium entry in endothelial cells. *The Journal of biological chemistry*, 282(22):16631–43, June 2007.

Ivan R Nabi. Cavin fever: regulating caveolae. 11(7):789–791, 2009.

Ivan R. Nabi and Phuong U. Le. Caveolae/raft-dependent endocytosis. *The Journal of cell biology*, 161(4):673–7, 2003.

D. Needham and R. S. Nunn. Elastic deformation and failure of lipid bilayer membranes containing cholesterol. *Biophysical journal*, 58(4):997–1009, October 1990.

Takehiro Ogata, Tomomi Ueyama, Koji Isodono, Masashi Tagawa, Naofumi Takehara, Tsuneaki Kawashima, Koichiro Harada, Tomosaburo Takahashi,

Tetsuo Shioi, Hiroaki Matsubara, and Hidemasa Oh. MURC, a muscle-restricted coiled-coil protein that modulates the Rho/ROCK pathway, induces cardiac dysfunction and conduction disturbance. *Molecular and cellular biology*, 28(10):3424–36, May 2008.

Lidiya Orlichenko, Bing Huang, Eugene Krueger, and Mark a McNiven. Epithelial growth factor-induced phosphorylation of caveolin 1 at tyrosine 14 stimulates caveolae formation in epithelial cells. *The Journal of biological chemistry*, 281(8):4570–9, March 2006.

Unn Örtegren, Margareta Karlsson, Natascha Blazic, Maria Blomqvist, Fredrik H Nystrom, Johanna Gustavsson, Pam Fredman, and Peter Stra. Lipids and glycosphingolipids in caveolae and surrounding plasma membrane of primary rat adipocytes. *European Journal of Biochemistry*, 271: 2028–2036, 2004.

George E. Palade. An Electron Microscope Study of the Mitochondrial Structure. *Structure*, 1953.

Ewa Paluch, Cécile Sykes, Jacques Prost, and Michel Bornens. Dynamic modes of the cortical actomyosin gel during cell locomotion and division. *Trends in cell biology*, 16(1):5–10, January 2006.

H. Park, Y. M. Go, R. Darji, J. W. Choi, M. P. Lisanti, M. C. Maland, and H. Jo. Caveolin-1 regulates shear stress-dependent activation of extracellular signal-regulated kinase. *American journal of physiology. Heart and circulatory physiology*, 278(4):H1285–93, April 2000.

S Park, D Koch, R Cardenas, J Kas, and C K Shih. Cell motility and local viscoelasticity of fibroblasts. *Biophysical journal*, 89(6):4330–42, December 2005.

Robert G. Parton and Kai Simons. The multiple faces of caveolae. *Nature reviews. Molecular Cell Biology*, 8(3):185–94, 2007.

Robert G. Parton, Michael Hanzal-Bayer, and John F. Hancock. Biogenesis of caveolae: a structural model for caveolin-induced domain formation. *Journal of cell science*, 119(Pt 5):787–96, March 2006.

Hemal H. Patel, Fiona Murray, and Paul A. Insel. Caveolae as organizers of pharmacologically relevant signal transduction molecules. *Annual review of pharmacology and toxicology*, 48:359–91, 2008.

J. Pécréaux, H.-G. Döbereiner, J. Prost, J.-F. Joanny, and P. Bassereau. Refined contour analysis of giant unilamellar vesicles. *The European physical journal. E, Soft matter*, 13(3):277–90, March 2004.

L. Pelkmans, J. Kartenbeck, and A. Helenius. Caveolar endocytosis of simian virus 40 reveals a new two-step vesicular-transport pathway to the ER. *Nature cell biology*, 3(5):473–83, May 2001.

Lucas Pelkmans and Marino Zerial. Kinase-regulated quantal assemblies and kiss-and-run recycling of caveolae. *Nature*, 436(7047):128–33, July 2005.

Basil J Petrof, J B Shrager, H H Stedman, A M Kelly, and H L Sweeney. Dystrophin protects the sarcolemma from stresses developed during muscle contraction. *Proceedings of the National Academy of Sciences of the United States of America*, 90(8):3710–4, April 1993.

Horst Pick, Evelyne L Schmid, Ana-paula Tairi, Erwin Ilegems, Ruud Hovius, and Horst Vogel. Investigating cellular signaling reactions in single attoliter vesicles. *Journal of the American Chemical Society*, 127(9):2908–12, March 2005.

Linda J. Pike, Xianlin Han, Koong-nah Chung, and Richard W Gross. Lipid Rafts Are Enriched in Arachidonic Acid and Plasmenylethanolamine and Their Composition Is Independent of Caveolin-1 Expression: A Quantitative Electrospray Ionization/Mass Spectrometric Analysis. *Biochemistry*, 41(6):2075–2088, February 2002.

Fabien Pinaud, Xavier Michalet, Gopal Iyer, Emmanuel Margeat, Hsiao-Ping Moore, and Shimon Weiss. Dynamic partitioning of a glycosyl-phosphatidylinositol-anchored protein in glycosphingolipid-rich microdomains imaged by single-quantum dot tracking. *Traffic (Copenhagen, Denmark)*, 10(6):691–712, June 2009.

Albert Pol, Sally Martin, Manuel A Fernández, Mercedes Ingelmo-Torres, Charles Ferguson, Carlos Enrich, and Robert G Parton. Cholesterol and fatty acids regulate dynamic caveolin trafficking through the Golgi complex and between the cell surface and lipid bodies. *Molecular biology of the cell*, 16(4):2091–105, April 2005.

Thomas D Pollard and John a Cooper. Actin, a central player in cell shape and movement. *Science (New York, N.Y.)*, 326(5957):1208–12, November 2009.

Thomas D. Pollard, Laurent Blanchoin, and R. Dyche Mullins. Molecular mechanisms controlling actin filament dynamics in nonmuscle cells. *Annual review of biophysics and biomolecular structure*, 29:545–76, January 2000.

Thomas Powers, Greg Huber, and Raymond Goldstein. Fluid-membrane tethers: Minimal surfaces and elastic boundary layers. *Physical Review E*, 65(4):41901., March 2002.

Andrew F. G. Quest, Jorge L. Gutierrez-Pajares, and Vicente A. Torres. Caveolin-1: an ambiguous partner in cell signalling and cancer. *Journal of cellular and molecular medicine*, 12(4):1130–50, August 2008.

Robert M. Raphael and Richard E. Waugh. Accelerated Interleaflet Transport of Phosphatidylcholine Membranes in Membranes Under Deformation. *Biophysical Journal*, 71:1374–1388, 1996.

D. Raucher and M. P. Sheetz. Membrane expansion increases endocytosis rate during mitosis. *J. Cell Biol.*, 144:497–506, 1999.

D. Raucher, T. Stauffer, W. Chen, K. Shen, S. Guo, J. D. York, M. P. Sheetz, and T. Meyer. Phosphatidylinositol 4, 5-bisphosphate functions as a second messenger that regulates cytoskeleton-plasma membrane adhesion. *Cell*, 100(2):221–228, 2000.

W. Rawicz, D. Needham, K. Olbrich, T. J. McIntosh, and E. Evans. Effect of chain length and unsaturation on elasticity of lipid bilayers. *Biophysical Journal*, 79(1):328–339, 2000a.

W. Rawicz, K. C. Olbrich, T. McIntosh, D. Needham, and E. Evans. Effect of chain length and unsaturation on elasticity of lipid bilayers. *Biophysical journal*, 79(1):328–39, July 2000b.

W. Rawicz, B. Smith, T. J. Mcintosh, S. Simon, and E. Evans. Elasticity, strength, and water permeability of bilayers that contain raft microdomain-forming lipids. *Biophysical Journal*, 94(12):4725–4736, 2008.

Babak Razani and Michael P. Lisanti. Caveolin-deficient mice: insights into caveolar function and human disease. *Journal of Clinical Investigation*, 108(11):1553–1561, December 2001.

Babak Razani, Scott E. Woodman, and Michael P. Lisanti. Caveolae: from cell biology to animal physiology. *Pharmacological reviews*, 54(3):431–67, September 2002.

Julia Riedl, Alvaro H Crevenna, Kai Kessenbrock, Jerry Haochen Yu, Dorothee Neukirchen, Michal Bista, Frank Bradke, Dieter Jenne, Tad A Holak, Zena Werb, Michael Sixt, and Roland Wedlich-Soldner. Lifeact: a versatile marker to visualize F-actin. *Nature methods*, 5(7):605–7, July 2008.

Victor Rizzo, Christine Morton, Natacha DePaola, Jan E Schnitzer, and Peter F Davies. Recruitment of endothelial caveolae into mechanotransduction pathways by flow conditioning in vitro. *American journal of physiology. Heart and circulatory physiology*, 285(4):H1720–9, 2003.

K. L. Rock, C. Gramm, L. Rothstein, K. Clark, R. Stein, L. Dick, D. Hwang, and A. L. Goldberg. Inhibitors of the proteasome block the degradation of most cell proteins and the generation of peptides presented on MHC class I molecules. *Cell*, 78(5):761–71, September 1994.

Winfried Römer, Ludwig Berland, Katharina Gaus, Barbara Windschiegl, Mohamed R E Aly, Vincent Fraisier, Jean-claude Florent, David Perrais, Christophe Lamaze, Claudia Steinem, Pierre Sens, Patricia Bassereau, and Ludger Johannes. SHIGA TOXIN INDUCES TUBULAR MEMBRANE INVAGINATIONS FOR ITS UPTAKE INTO CELLS. *Cell*, 2008.

Paul E. Ross, Sarah S. Garber, and Micheal D. Cahalan. Membrane chloride conductance and capacitance in Jurkat T lymphocytes during osmotic swelling. *Biophysical journal*, 66(1):169–78, January 1994.

Olivier Rossier, D. Cuvelier, N. Borghi, P. H. Puech, I. Derényi, A. Buguin, P. Nassoy, and F. Brochard-Wyart. Giant Vesicles under Flows: Extrusion and Retraction of Tubes. *Langmuir*, 19(3):575–584, February 2003.

K. G. Rothberg, J. E. Heuser, W. C. Donzell, Y. S. Ying, J. R. Glenney, and R. G. Anderson. Caveolin, a protein component of caveolae membrane coats. *Cell*, 68(4):673–82, February 1992.

C. Rotsch and M. Radmacher. Drug-induced changes of cytoskeletal structure and mechanics in fibroblasts: an atomic force microscopy study. *Biophysical Journal*, 78:520–535, 2000.

Aurélien Roux, Giovanni Cappello, Jean Cartaud, Jacques Prost, Bruno Goud, and Patricia Bassereau. A minimal system allowing tubulation with molecular motors pulling on giant liposomes. *Proceedings of the National Academy of Sciences of the United States of America*, 99(8):5394–9, April 2002.

V. O. Rybin, X. Xu, M. P. Lisanti, and S. F. Steinberg. Differential targeting of beta -adrenergic receptor subtypes and adenylyl cyclase to cardiomyocyte caveolae. A mechanism to functionally regulate the cAMP signaling pathway. *The Journal of biological chemistry*, 275(52):41447–57, December 2000.

O. Sandre, L. Moreaux, and F. Brochard-Wyart. Dynamics of transient pores in stretched vesicles. *Proceedings of the National Academy of Sciences of the United States of America*, 96(19):10591–6, September 1999.

P. E. Scherer, T. Okamoto, M. Chun, I. Nishimoto, H. F. Lodish, and M. P. Lisanti. Identification, sequence, and expression of caveolin-2 defines a caveolin gene family. *Proc. Natl. Acad. Sci. USA*, 93:131–135, 1996.

M. Schliwa. Action of cytochalasin D on cytoskeletal networks. *The Journal of Cell Biology*, 92(1):79, 1982.

R. E. Scott. Plasma membrane vesiculation: a new technique for isolation of plasma membranes. *Science (New York, N.Y.)*, 194(4266):743–5, November 1976.

Daniel G. Sedding, Jennifer Hermsen, Ulrike Seay, Oliver Eickelberg, Wolfgang Kummer, Carsten Schwencke, Ruth H. Strasser, Harald Tillmanns, and Ruediger C. Braun-Dullaeus. Caveolin-1 facilitates mechanosensitive protein kinase B (Akt) signaling in vitro and in vivo. *Circulation research*, 96(6):635–42, 2005.

Adrian M. J. Senderowicz, Gurmeet Kaur, Eduardo Sainz, Charmaine Laing, Wayne D. Inman, Jaime Odriguez, Phillip Crews, Louis Malspeis, Michael R. Grever, Edward A. Sausville, and Kimberly L.K. Duncan. Jasplakinolide's Inhibition of the Growth of Prostate Carcinoma Cells In Vitro With Disruption of the Actin Cytoskeleton. *Journal of the National Cancer Institute*, 87(1), 1995.

Pierre Sens and Matthew Turner. Budded membrane microdomains as regulators for cellular tension. pages 1–4, 2005.

Michael P. Sheetz. Cell control by membrane-cytoskeleton adhesion. *Nature reviews. Molecular cell biology*, 2(5):392–6, May 2001.

Michael P. Sheetz, Julia E. Sable, and Hans-Günther Döbereiner. Continuous Membrane-Cytoskeleton Adhesion Requires Continuous Accommodation to Lipid and Cytoskeleton Dynamics. *Annual Reviews Biophysical and Biomolecular Structures*, 35:417–434, 2006.

Piotr Sicinski, Y Geng, A S Ryder-Cook, Eric A Barnard, Mark G Darlison, and P J Barnard. The molecular basis of muscular dystrophy in the mdx mouse: a point mutation. *Science (New York, N.Y.)*, 244(4912):1578–80, June 1989.

K. Simons and D. Toomre. Lipid rafts and signal transduction. *Nature Reviews. Molecular Cell Biology*, 1(1):31–39, 2000.

C. Solsona, B. Innocenti, and J. M. Fernández. Regulation of exocytotic fusion by cell inflation. *Biophysical journal*, 74(2 Pt 1):1061–73, February 1998.

Kenneth S. Song, Philipp E. Scherer, ZhaoLan Tang, Takashi Okamoto, Shengwen Li, Mark Chafel, Caryn Chu, D. Stave Kohtz, and Michael P. Lisanti. Expression of Caveolin-3 in Skeletal, Cardiac, and Smooth Muscle Cells. *Biochemistry*, 271(25):15160 –15165, 1996.

Kenneth S. Song, ZhaoLan Tang, Shengwen Li, and Michael P. Lisanti. Mutational Analysis of the Properties of Caveolin-1. *Biochemistry*, 272(7): 4398 –4403, 1997.

Benoit Sorre. *Role of Membrane Curvature in Intracellular Trafficking*. PhD thesis, 2010.

Benoit Sorre, Andrew Callan-Jones, Jean-Baptiste Manneville, Pierre Nassoy, Jean-François Joanny, Jacques Prost, Bruno Goud, and Patricia Bassereau. Curvature-driven lipid sorting needs proximity to a demixing point and is aided by proteins. *Proceedings of the National Academy of Sciences of the United States of America*, 106(14):5622–6, April 2009.

Federica Sotgia, Scott E Woodman, Gloria Bonuccelli, Franco Capozza, Carlo Minetti, Philipp E Scherer, and Michael P Lisanti. Phenotypic behavior of caveolin-3 R26Q, a mutant associated with hyperCKemia, distal myopathy, and rippling muscle disease. *American journal of physiology. Cell physiology*, 285(5):C1150–60, 2003.

Dimitrije Stamenovic and Ning Wang. Cellular Responses to Mechanical Stress Invited Review: Engineering approaches to cytoskeletal mechanics. *Journal of Applied Physiology*, 89:2085–2090, 2000.

Sergei V. Strelkov, Harald Herrmann, and Ueli Aebi. Molecular architecture of intermediate filaments. *BioEssays : news and reviews in molecular, cellular and developmental biology*, 25(3):243–51, March 2003.

Maria Sverdlov, Ayesha N Shajahan, and Richard D Minshall. Tyrosine phosphorylation-dependence of caveolae-mediated endocytosis. *Journal of cellular and molecular medicine*, 11(6):1239–50, 2007.

S. Svetina, B. Zeks, R. E. Waugh, and R. M. Raphael. Theoretical analysis of the effect of the transbilayer movement of phospholipid molecules on the dynamic behavior of a microtube pulled out of an aspirated vesicle. *European biophysics journal : EBJ*, 27(3):197–209, January 1998.

K. Svoboda and S. M. Block. Biological applications of optical forces. *Annual review of biophysics and biomolecular structure*, 23:247–85, January 1994.

A. Tagawa, A. Mezzacasa, A. Hayer, A. Longatti, L. Pelkmans, and A. Helenius. Assembly and trafficking of caveolar domains in the cell: caveolae as stable, cargo-triggered, vesicular transporters. *J Cell Biol*, 170:769–79, 2005.

Michael J. Taggart. Smooth muscle excitation-contraction coupling: a role for caveolae and caveolins? *News in physiological sciences : an international journal of physiology produced jointly by the International Union of*

Physiological Sciences and the American Physiological Society, 16(April): 61–5, April 2001.

Peter Thomsen, Kirstine Roepstorff, Martin Stahlhut, and Bo Van Deurs. Caveolae Are Highly Immobile Plasma Membrane Microdomains, Which Are not Involved in. *Molecular Biology of the Cell*, 13(January):238 –250, 2002.

Hans Thorn, Karin G Stenkula, Margareta Karlsson, Unn Ortegren, Fredrik H Nystrom, Johanna Gustavsson, and Peter Stralfors. Cell surface orifices of caveolae and localization of caveolin to the necks of caveolae in adipocytes. *Molecular biology of the cell*, 14(10):3967–76, October 2003.

Aiwei Tian, Corinne Johnson, Wendy Wang, and Tobias Baumgart. Line Tension at Fluid Membrane Domain Boundaries Measured by Micropipette Aspiration. *Physical Review Letters*, 98(20):18–21, May 2007.

Dominique Trouet, Bernd Nilius, Axel Jacobs, Claude Remacle, Guy Droogmans, and Jan Eggermont. Caveolin-1 modulates the activity of the volume-regulated chloride channel. *The Journal of physiology*, 520 Pt 1: 113–9, October 1999.

P. L. Vaghy, Jin Fang, Wenrong Wu, and Laszlo P. Vaghy. Increased caveolin-3 levels in mdx mouse muscles. *FEBS letters*, 431(1):125–7, July 1998.

J. Vinten, M. Voldstedlund, H. Clausen, K. Christiansen, J. Carlsen, and J. Tranum-Jensen. A 60-kDa protein abundant in adipocyte caveolae. *Cell and Tissue Research*, 305(1):99–106, July 2001.

J. Vinten, A. H. Johnsen, P. Roepstorff, J. Harpø th, and J. Tranum-Jensen. Identification of a major protein on the cytosolic face of caveolae. *Biochimica et biophysica acta*, 1717(1):34–40, November 2005.

Ulla Vogel, Kirsten Sandvig, and Bo Van Deurs. Expression of caveolin-1 and polarized formation of invaginated caveolae in Caco-2 and MDCK II cells. *Journal of Cell Science*, 111:825–832, 1998.

Viola Vogel and Michael Sheetz. Local force and geometry sensing regulate cell functions. *Nature reviews. Molecular Cell Biology*, 7(4):265–75, April 2006.

M. Voldstedlund, J. Vinten, and J. Tranum-Jensen. cav-p60 expression in rat muscle tissues. Distribution of caveolar proteins. *Cell and tissue research*, 306(2):265–76, November 2001.

T. Wakatsuki, B. Schwab, NC Thompson, and EL Elson. Effects of cytochalasin D and latrunculin B on mechanical properties of cells. *Journal of Cell Science*, 114(5):1025, 2001.

Gregory Q Wallace and Elizabeth M McNally. Mechanisms of muscle degeneration, regeneration, and repair in the muscular dystrophies. *Annual review of physiology*, 71:37–57, January 2009.

R. Waugh and E. Evans. Thermoelasticity of red blood cell. *Biophysical Journal*, 26(April):115–131, 1979.

Richard E Waugh. SURFACE VISCOSITY MEASUREMENTS FROM

LARGE BILAYER VESICLE TETHER FORMATION II. Experiments. *Biophys J*, 38:29–37, 1982.

M Way and R G Parton. M-caveolin, a muscle-specific caveolin-related protein. *FEBS Lett*, 376:108–12, 1995.

E Yamada. The fine structure of the renal glomerulus of the mouse. *The journal of histochemistry and cytochemistry : official journal of the Histochemistry Society*, 3(4):309, 1955.

Daniel L. Yamamoto, Robert I. Csikasz, Yu Li, Gunjana Sharma, Klas Hjort, Roger Karlsson, and Tore Bengtsson. Myotube formation on micropatterned glass: intracellular organization and protein distribution in C2C12 skeletal muscle cells. *The journal of histochemistry and cytochemistry : official journal of the Histochemistry Society*, 56(10):881–92, October 2008.

Joshua Zimmerberg and Michael M. Kozlov. How proteins produce cellular membrane curvature. *Nature reviews. Molecular Cell Biology*, 7(1):9–19, January 2006.

Die VDM Verlagsservicegesellschaft sucht für wissenschaftliche Verlage abgeschlossene und herausragende

Dissertationen, Habilitationen, Diplomarbeiten, Master Theses, Magisterarbeiten usw.

für die kostenlose Publikation als Fachbuch.

Sie verfügen über eine Arbeit, die hohen inhaltlichen und formalen Ansprüchen genügt, und haben Interesse an einer honorarvergüteten Publikation?

Dann senden Sie bitte erste Informationen über sich und Ihre Arbeit per Email an *info@vdm-vsg.de*.

Sie erhalten kurzfristig unser Feedback!

VDM Verlagsservicegesellschaft mbH
Dudweiler Landstr. 99   Telefon +49 681 3720 174
D - 66123 Saarbrücken   Fax   +49 681 3720 1749
www.vdm-vsg.de

Die VDM Verlagsservicegesellschaft mbH vertritt

Printed by Books on Demand GmbH, Norderstedt / Germany